Information Circular 9512

Proceedings of the International Workshop o
Numerical Modeling for Underground Mine Exc sign

Edited by Gabriel S. Esterhuizen, Ph.D., Christopher Mark, Ph.D., P.E., Ted M. Klemetti, and Robert J. Tuchman

DEPARTMENT OF HEALTH AND HUMAN SERVICES
Centers for Disease Control and Prevention
National Institute for Occupational Safety and Health
Pittsburgh Research Laboratory
Pittsburgh, PA

June 2009

Disclaimer

Mention of any company or product does not constitute endorsement by the National Institute for Occupational Safety and Health (NIOSH). In addition, citations to Web sites external to NIOSH do not constitute NIOSH endorsement of the sponsoring organizations or their programs or products. Furthermore, NIOSH is not responsible for the content of these Web sites. All web addresses referenced in this document were accessible as of the publication date.

The views expressed by non-NIOSH authors in these proceedings are not necessarily those of NIOSH.

Ordering Information

To receive documents or other information about occupational safety and health topics, contact NIOSH at

Telephone: **1–800–CDC–INFO** (1–800–232–4636)
TTY: 1–888–232–6348
e-mail: cdcinfo@cdc.gov

or visit the NIOSH Web site at **www.cdc.gov/niosh**.

For a monthly update on news at NIOSH, subscribe to NIOSH *eNews* by visiting **www.cdc.gov/niosh/eNews**.

DHHS (NIOSH) Publication No. 2009–141

June 2009

SAFER • HEALTHIER • PEOPLE™

CONTENTS

UNIT OF MEASURE ABBREVIATIONS USED IN THIS REPORT

ft	foot
GPa	gigapascal
in	inch
kg/m^3	kilogram per cubic meter
km	kilometer
km^2	square kilometer
kPa	kilopascal
L/s	liter per second
lb/ft^3	pound per cubic foot
m	meter
m^2	square meter
m^3	cubic meter
m^3/t	cubic meter per ton
m/sec	meter per second
md	millidarcy
mm	millimeter
MPa	megapascal
Mt	million tons
Mtpa	million tons per annum
Pa	pascal
psi	pound-force per square inch
tpd	tons per day

PROCEEDINGS OF THE INTERNATIONAL WORKSHOP ON NUMERICAL MODELING FOR UNDERGROUND MINE EXCAVATION DESIGN

Edited by Gabriel S. Esterhuizen, Ph.D.,[1] Christopher Mark, Ph.D., P.E.,[2] Ted M. Klemetti,[3] and Robert J. Tuchman[4]

ABSTRACT

Numerical models play a significant role in the design of safe underground mining excavations and support systems. Advances in the capabilities of numerical modeling software, together with ever increasing computational speeds, have made it possible to investigate the very nature of the large-scale rock mass and its response to mining excavations. The improved understanding of the rock response obtained from modeling enhances our designs, resulting in greater stability and safety of the mining excavations. To help advance the state of the art in this field, the National Institute for Occupational Safety and Health organized the International Workshop on Numerical Modeling for Underground Mine Excavation Design. The workshop was held in Asheville, NC, on June 28, 2009, in association with the 43rd U.S. Rock Mechanics Symposium. The proceedings include 10 papers from leading rock mechanics and numerical modeling experts in the United States, Canada, Australia, and Germany. The papers address a wide range of issues, including various numerical modeling approaches, rock mass modeling, and applications in coal and metal mines.

---

[1]Senior research engineer, Pittsburgh Research Laboratory, National Institute for Occupational Safety and Health, Pittsburgh, PA.
[2]Principal research engineer, Pittsburgh Research Laboratory, National Institute for Occupational Safety and Health, Pittsburgh, PA.
[3]Research engineer, Pittsburgh Research Laboratory, National Institute for Occupational Safety and Health, Pittsburgh, PA.
[4]Technical writer-editor, Writer-Editor Services Branch, Division of Creative Services, National Center for Health Marketing, Coordinating Center for Health Information and Service, Centers for Disease Control and Prevention, Pittsburgh, PA.

AN EFFICIENT APPROACH TO NUMERICAL SIMULATION OF COAL MINE-RELATED GEOTECHNICAL ISSUES

By D. P. Adhikary, Ph.D.,[1] and H. Guo, Ph.D.[1]

ABSTRACT

Reliable prediction of mine stability, surface subsidence, mine water inflow, and mine gas emissions is essential not only for improving mine safety and reducing coal production costs, but also for assessing and managing the environmental impact of mining.

This paper describes an integrated approach to simulation and prediction of mining-induced surface subsidence, mine groundwater inflow, aquifer interference, and mine gas emission. It involves a combination of site geological, geotechnical, and hydrogeological characterization; study of surface subsidence and subsurface rock caving mechanisms; monitoring of pore pressure changes of the surrounding strata, mine water inflows, and mine gas emission; and three-dimensional (3-D) numerical modeling. Central to this integrated approach is a 3-D computer code called COSFLOW developed by CSIRO Exploration and Mining of Australia in collaboration with NEDO and JCOAL of Japan to address the coal mine-related issues. COSFLOW incorporates unique features (e.g., Cosserat continuum formulation) that make it ideal for simulating coal mining-related issues and examining the interaction between rock fracture, aquifer interference and water flow, and gas emission.

INTRODUCTION

Mine subsidence is becoming a major issue of community concern. Mining-induced subsidence can significantly affect mining costs where major surface facilities and natural environments need to be protected. Longwall mining has been undertaken under important surface structures such as river systems, gorges, cliffs, power lines, pipelines, communication cables, major roads, and bridges.

Similarly, the ability to accurately predict the behavior of water in longwall operations has become a pressing issue. While higher-capacity longwall mines are putting pressure on dewatering requirements, environmental concerns about loss of water supply overshadow many mines. In the future, environmental water issues, associated with major aquifers, could affect a mine's ability to gain mining approval.

As the mines are getting deeper and gassier, prediction and management of mine gas-related issues are becoming increasingly important.

Over the last 6 years, CSIRO Exploration and Mining has been involved in developing an integrated approach to predict the impact of mining on surface subsidence, subsurface aquifers, mine water inflow into underground coal mines, and mine gas emission. It involves a combination of site geotechnical and hydrogeological characterization, study of surface subsidence and subsurface rock caving mechanisms, monitoring of pore pressure changes of the surrounding strata, mine water inflows, mine gas emission, and 3-D modeling.

This paper introduces the integrated approach and the 3-D finite-element code called COSFLOW, which was developed to address longwall coal mine-related issues. COSFLOW was developed as a result of a major joint project between CSIRO (Australia) and NEDO and JCOAL (Japan).

A unique feature of COSFLOW is the incorporation of Cosserat continuum theory [Adhikary and Dyskin 1997] in its formulation. In the Cosserat model, interlayer interfaces (joints, bedding planes) are considered to be smeared across the mass, i.e., the effects of interfaces are incorporated implicitly in the choice of stress-strain model formulation. An important feature of the Cosserat model is that it incorporates bending rigidity of individual layers in its formulation, which makes it different (and more efficient in simulating stratified strata) from other conventional implicit models.

SITE CHARACTERIZATION

The CSIRO approach to assessing the impact of mining on subsurface aquifers and mine water inflow into underground coal mines involves a combination of 3-D geological, geotechnical, and hydrogeological site characterization, study of surface subsidence and subsurface rock caving mechanisms, monitoring of pore pressure changes of the surrounding strata and underground water inflows, mine gas emissions, and 3-D numerical modeling. The work carried out is described in Guo et al. [2002, 2003a] and Adhikary et al. [2004]. This section briefly describes the site characterization.

The 3-D site geotechnical and hydrogeological characterization typically includes:

(1) Collection of available regional and site-specific hydrogeological data;

[1]CSIRO Exploration and Mining, Australia.

(2) Dedicated drilling program with geological and geophysical logging for all drillholes;
(3) In situ stress and strata permeability measurements;
(4) Laboratory testing of rock and coal strengths;
(5) Automatic drillhole geotechnical interpretation;
(6) Two-dimensional (2-D) and 3-D seismic surveys (where applicable);
(7) Field geotechnical and hydrogeological monitoring (surface piezometric and extensometer monitoring, and underground water inflow (gas emission) measurements); and
(8) In situ 3-D geological, geotechnical, hydrogeological, and gas models.

CSIRO has been actively involved in such site characterization studies, including planning, organization, testing, data interpretation, and preparation of a site hydrogeological and geotechnical conceptual models. Figure 1 shows an example of a modeled stratigraphy from a deep mining study at a site in New South Wales, Australia [Guo et al. 2003b].

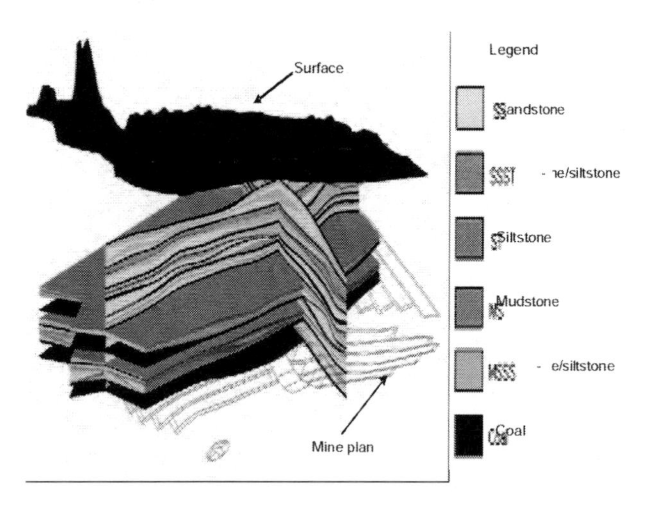

Figure 1.—Perspective view of 3-D geotechnical model of a mine site in New South Wales, Australia, looking from the southwest (an area of 5 km by 6 km modeled with depth ranging from 200 to 600 m; colors represent various strata units).

For the past 5 years, CSIRO has conducted site monitoring, data interpretation, and numerical modeling work at Springvale Mine to predict water inflow into its longwall panels. Springvale currently mines the 12th longwall block, extracting the lower 3.2-m section of the 7-m Lidsdale/Lithgow Seam, 360 m below the Newnes Plateau. At an average production rate of 2.75 Mtpa, the projected mine life is 25 years [Knight and Miller 2005].

As part of site hydrogeological characterization at the Springvale Mine, CSIRO analyzed the geophysical drillhole data. Density, natural gamma, and sonic velocity logs were interpreted using the CSIRO/CMTE computer program LogTrans to obtain detailed strata classification. The interpretation involved the following key steps:

(1) Detailed analysis of key control cored holes where geological logs and core photographs were available;
(2) Statistical characterization of the geophysical signatures of these geological units using LogTrans; and
(3) Automatic computer interpretation of geophysical logs of other open holes for geological units using the identified geophysical signatures.

A verification (or training) of LogTrans against the existing geological data is the first step in automated strata interpretation. A careful verification exercise was conducted using the geological and geophysical data from three control drillholes. In the second step, LogTrans was used to interpret the remaining drillholes. Figure 2 shows LogTrans interpretations of geophysical data from five drillholes.

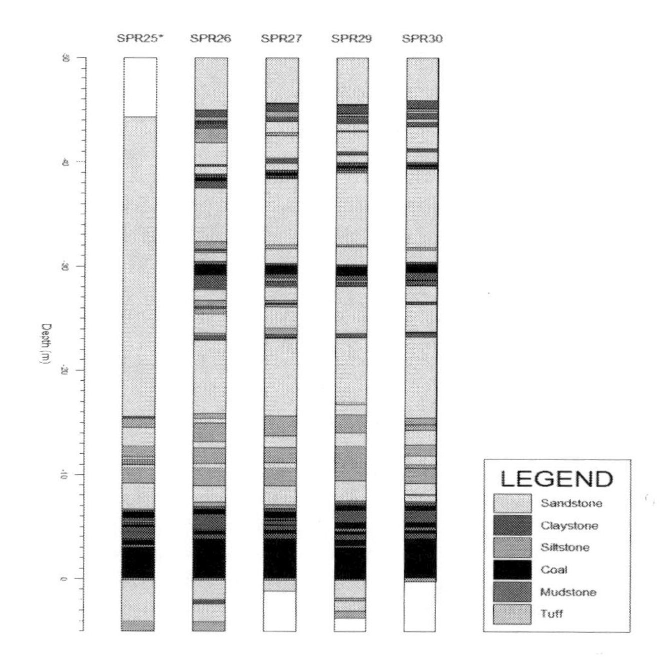

Figure 2.—LogTrans interpretation of Springvale Mine data, to 50 m above the mining seam (SPRxx denotes the Springvale borehole number) (after Guo et al. [2002]).

As an integral part of the site study, piezometric monitoring of the surrounding strata during mining was carried out to assess groundwater response to mining. A total of 26 piezometers were installed at different horizons in 8 boreholes (Figure 3). A typical groundwater pressure response is shown in Figure 4. Such a response verifies the existence of the aquifer, and the water pressure change also provides essential input data to the subsequent coupled numerical modeling with COSFLOW.

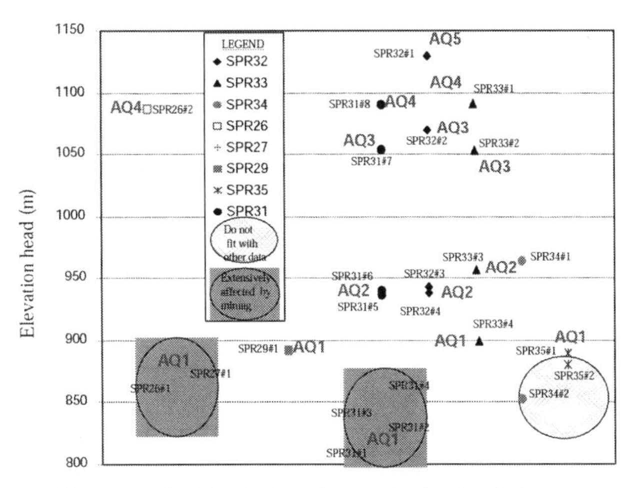

Figure 3.—Aquifer delineation on the basis of piezometer data (after Knight and Miller [2005]).

The next step in the mining impact assessment is to develop a conceptual hydrogeological model with proper delineation of aquifers and aquitards. On the basis of local geology and the piezometric results, CSIRO has developed a conceptual hydrogeological model (Figure 5) for the mine (see Knight and Miller [2005]). Five distinctive aquifers (AQ1 to AQ4) separated by low-permeability layers (aquitards) were identified.

LONGWALL MINING-INDUCED GROUND DEFORMATION

A number of researchers have investigated the mechanics of strata deformation induced by longwall mining. These studies have recognized four distinctive deformation zones in the overburden rock (Figure 6). Although

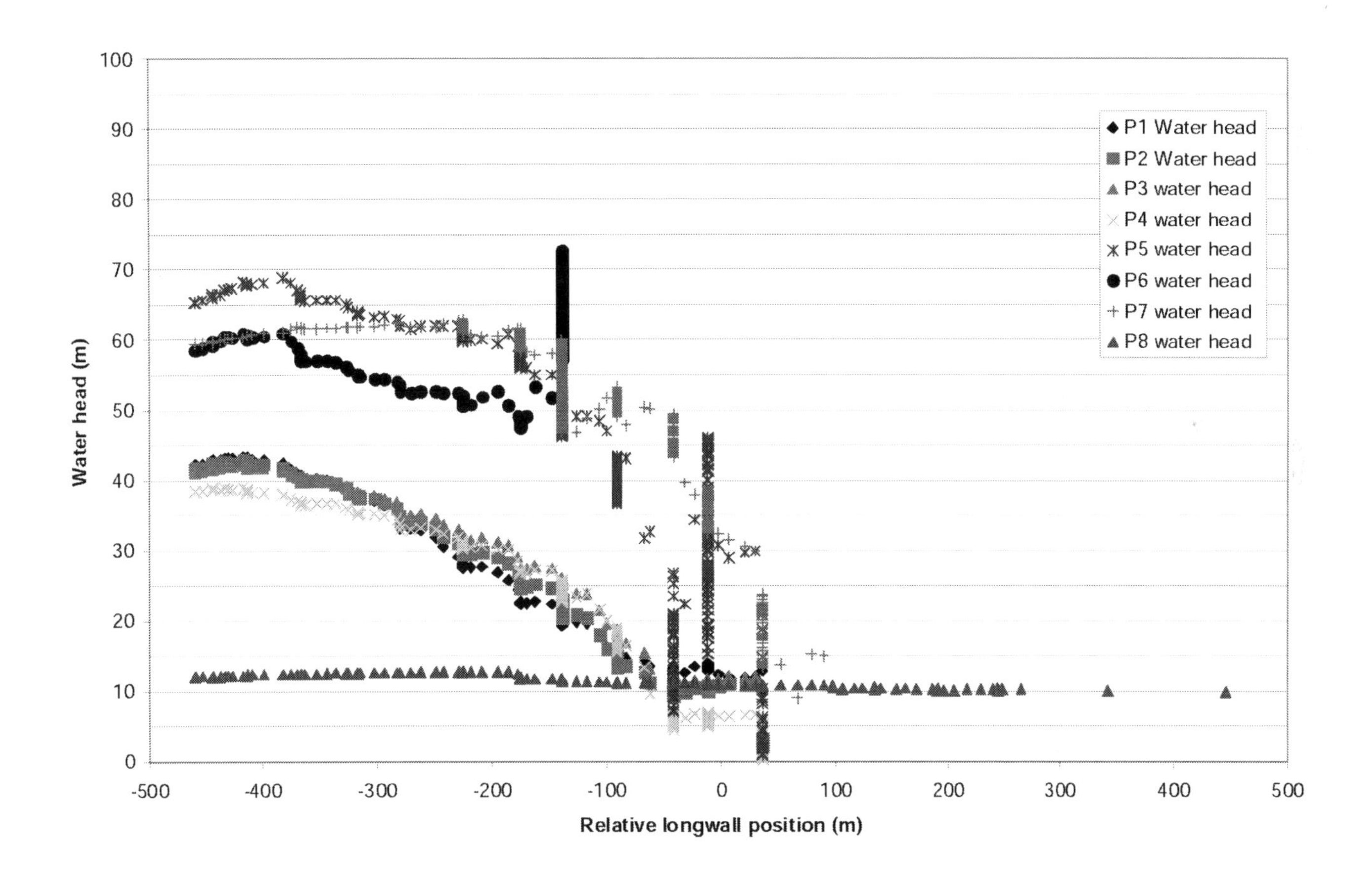

Figure 4.—Piezometer readings at SPR31; note the drop in piezometric pressures as the longwall face approached the borehole location (after Guo et al. [2006]).

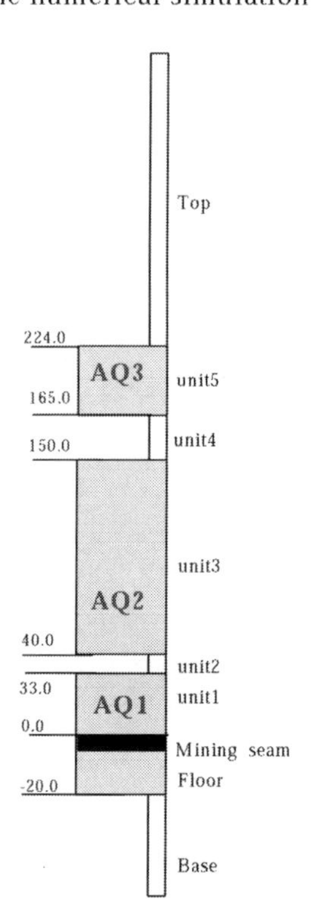

Figure 5.—A conceptual hydrogeological model for Springvale Mine (after Knight and Miller [2005]).

terminology is not precise, they can be typically described (in order of increasing height above the mining seam) as:

(1) A caving zone with broken blocks of rock detached from the roof (less than 10 times the extraction height, probably about 3–5 times the extraction height);
(2) A disturbed or fractured zone where the rocks have sagged downward and consequently suffered bending, fracturing, joint opening, and bed separation (about 15–40 times the extraction height);
(3) A constrained zone where the strata have sagged slightly over the panel without suffering significant fracturing or alteration to the original geomechanical properties (variable thickness); and
(4) A surface zone with tensile fracturing (up to 20 m thick).

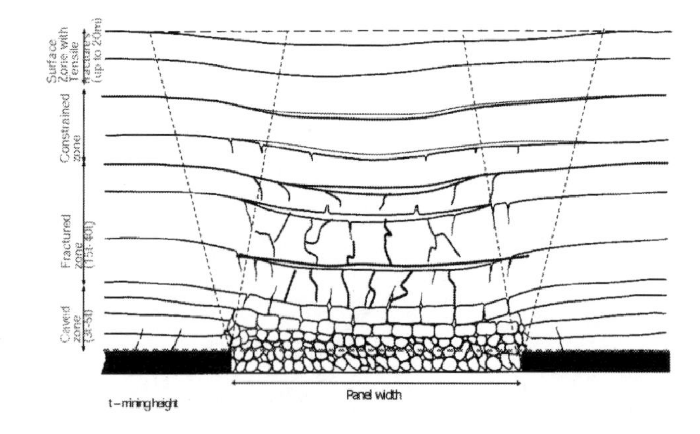

Figure 6.—Deformation zones induced by longwall mining.

The vertical extent of each of these zones is variable and depends on many factors (e.g., longwall width, extraction height, overburden rock properties, local geology, and depth of cover). The permeability of rock strata within the first two zones can significantly increase and, as a result, water from aquifers located within these two zones may be drawn into the mining panels.

COSFLOW FORMULATION

This section describes some important aspects of the 3-D finite-element code COSFLOW. Here we provide a loosely coupled mechanical and two-phase fluid flow formulation, where the mechanical equations and fluid flow equations are solved sequentially. Such an approach of solving a coupled mechanical and two-phase fluid flow problem is widely used in the scientific community [Rutqvist et al. 2002]. In such an approach, mechanical effects are assumed to occur almost instantaneously when compared to diffusion effects. In COSFLOW, sequential cycling between fluid flow steps and mechanical steps is controlled through a limiting parameter that would switch off the fluid flow steps and start the mechanical steps once the maximum change in average pore pressure within the model exceeds the value set by the limiting parameter.

A unique feature of COSFLOW is the incorporation of Cosserat continuum theory [Cosserat and Cosserat 1909] in its formulation. In the Cosserat model, interlayer interfaces (joints, bedding planes) are considered to be smeared across the mass, i.e., the effects of interfaces are incorporated implicitly in the choice of stress-strain model formulation. An important feature of the Cosserat model is that it incorporates bending rigidity of individual layers in its formulation, which makes it different from other conventional implicit models.

The flow of either phase of fluid is controlled by the permeability of the porous medium, which remains a highly nonlinear function of mining-induced stress and resulting fractures. Thus, in order to be able to correctly estimate water inflow or gas emission, it is important not only to estimate the initial permeability correctly, but also to compute its variation during mining. In this code, permeability change during mining is computed as a function of the mining-induced strain.

Mechanical (Cosserat) Model

Coal measure rocks are essentially stratified in nature. Since stratified rock masses exhibit highly anisotropic strength and deformation characteristics, it is necessary to include effects of stratification into the mathematical formulations describing the load deformation behavior of such rock masses.

There are essentially two different approaches that can be adopted in simulating the load deformation behavior of stratified rock masses. Firstly, the layered medium can be defined in a discontinuum manner (i.e., discrete modeling) such that each and every joint and intact layer is discretely defined in numerical simulation. However, when closely spaced joints occur in large numbers such that the layer thickness becomes much smaller than the dimensions of the problem region, the discrete modeling of such a medium becomes tedious and expensive to perform. The only feasible and elective way of incorporating the influence of such a system of joints in the analysis is to formulate an equivalent model within the framework of continuum mechanics with an appropriate constitutive relationship.

A continuum description of a layered medium can be formulated as long as consistency and statistical homogeneity in joint properties and spacing can be established. Such a continuum model provides a large-scale (average) description of the material response to loading. The continuum model devised in such a manner is often known as a smeared (implicit) joint model in the sense that the joints are implicit in the choice of the stress-strain relationship adopted for the equivalent continuum. A distinctive advantage of the smeared joint model is that in a numerical (e.g., finite-element) solution, the problem region can now be discretized with a coarser mesh (i.e., subdivided into fewer finite elements) than in the discrete models, where the size of the finite elements cannot exceed the layer thickness. Thus, in smeared joint models, the size of the elements is dictated solely by computational needs rather than by the layer thickness.

In the models based on the conventional equivalent continuum approach, the layered material is replaced with a homogeneous anisotropic medium characterized by the so-called effective elastic moduli comprising the heterogeneity of the medium. The methods of computing the effective characteristics for layered media were first proposed by Lifshitz and Rosenzweig [1946, 1951] for arbitrary anisotropic materials of the layers, then by Salamon [1968] for the case of transversely isotropic layers and Gerrard [1982] for the orthotropic layers. If sliding between the layers can occur during the deformation process, the equivalent continuum should be viewed as a continuum of elastoplastic type (e.g., Zienkiewicz and Pande [1977]; Gerrard and Pande [1985]; Alehossein and Carter [1990]). Such equivalent continuum models may provide reasonably accurate predictions when joint slips are minimal, i.e., the layer bending can be neglected. However, when the joint slips are large and the layers can bend as they slip against each other, models based on such conventional continuum models may considerably overestimate the deformation since the bending rigidity of the layers are not incorporated in the model formulation.

In the case of rock layers with bending stiffness, an equivalent continuum model can be formulated successfully on the basis of Cosserat theory [Adhikary and Guo 2002; Adhikary and Dyskin 1997, 1998]. Adhikary and Dyskin [1997] provide a thorough analysis and comparison between the conventional equivalent continuum and the Cosserat continuum models.

Whereas the conventional continuum model has 3 independent degrees of freedom and 6 independent stress components in a 3-D case, the Cosserat model for the stratified material will have 6 independent degrees of freedom and 10 independent stress components.

The layer interfaces can exhibit three different modes of behavior: (1) elastically connected with the interface normal and shear stiffness, (2) plastic with frictional sliding, and (3) disconnected with tensile opening. Similarly, the rock layer may either deform elastically or may sustain some plastic deformation as well. A description of the full elastoplastic Cosserat formulation is provided by Adhikary et al. [2002]. For simplicity, a 2-D plane strain Cosserat formulation will be presented in this paper. Using the Cartesian coordinates (x1,x2) in two dimensions, the material point displacement can be defined by a translational vector (u1,u2) and by a rotation Ω3. Here, x3 is aligned to the out-of-plane direction and x2 is perpendicular to the layers.

The general 2-D Cosserat model has four non-symmetric stress components σ_{11}, σ_{22}, σ_{21}, σ_{12} and two couple stresses m_{31}, m_{32}. When the rock layers are aligned in the x_1 direction, the moment stress term m_{32} vanishes. The four stresses are conjugate to four deformation measures γ_{11}, γ_{22}, γ_{21}, γ_{12} defined by:

$$\gamma_{ij} = \frac{\partial u_j}{\partial x_i} - \varepsilon_{3ij}\Omega_3 \qquad (1)$$

where i and j equal 1 or 2 and ε_{3ij} is the permutation tensor ($\varepsilon_{312} = -\varepsilon_{321} = 1$ and $\varepsilon_{311} = \varepsilon_{322} = 0$). The couple stress m_{31} is conjugate to the respective curvature κ_1 defined by:

$$\kappa_1 = \frac{\partial \Omega_3}{\partial x_1} \qquad (2)$$

The elastic stress strain relationships are described by:

$$\sigma = [D_e]e_e \qquad (3)$$

where

$$\sigma = \{\sigma_{11}, \sigma_{22}, \sigma_{21}, \sigma_{12}, \mathrm{m}_{31}\}$$
$$e = \{\gamma_{11}, \gamma_{22}, \gamma_{21}, \gamma_{12}, \kappa_1\} \qquad (4)$$

$$D = \begin{bmatrix} A_{11} & A_{12} & 0 & 0 & 0 \\ & A_{22} & 0 & 0 & 0 \\ & & G_{11} & G_{12} & 0 \\ & symm & & G_{22} & 0 \\ & & & & B_1 \end{bmatrix} \qquad (5)$$

where

$$A_{11} = \frac{E}{1-\nu^2 - \dfrac{\nu^2(1+\nu)^2}{1-\nu^2 + \dfrac{E}{hk_n}}} \qquad (6)$$

$$A_{22} = \frac{1}{\dfrac{1-\nu-2\nu^2}{E(1-\nu)} + \dfrac{1}{hk_n}} \qquad (7)$$

$$A_{12} = \frac{\nu}{1-\nu}A_{22} \qquad (8)$$

$$\frac{1}{G_{11}} = \frac{1}{G} + \frac{1}{hk_s} \qquad (9)$$

$$G_{11} = G_{12} = G_{21} \qquad (10)$$

$$G_{22} = G_{11} + G \qquad (11)$$

$$B_1 = \frac{Eh^2}{12(1-\nu^2)}\left(\frac{G-G_{11}}{G+G_{11}}\right) \qquad (12)$$

and E is the Young's modulus of the intact layer, ν is the Poisson's ratio, h is the layer thickness, G is the shear modulus of the intact layer, and k_n and k_s are the joint normal and shear stiffnesses.

Flow Model

In COSFLOW, a porous medium is simulated as a region having two porosities—one representing a continuum porous rock (primary porosity), the other a fracture network (secondary porosity). Thus, the flow behavior is described mainly by the interaction of the basic components, namely, the porous matrix and the surrounding fracture system. The fractures provide rapid hydraulic connection but little fluid mass storage, whereas the porous matrix represents high storage but low hydraulic connection. The flow model incorporated in COSFLOW is similar to the conventional flow model, i.e., the flow in the fracture (cleat) system is controlled by the pressure gradient and is described using Darcy's law, whereas the desorption (flow in the matrix) is controlled by the concentration gradient and is described using Fick's law (e.g., King and Ertekin [1989a,b; 1991; 1994; 1995] provide a comprehensive list of references and surveys of mathematical models used in the simulation of gas production from coal seams). The relationship between gas concentration and pressure is a nonlinear function and is described using Langmuir [1916] equations.

The flow model adopted in this study can be briefly described in the following manner. By assuming that the flow of fluid (gas/water) obeys Darcy's law, the continuity requirement of each fluid phase can be expressed through the following sets of equations:

$$\nabla \cdot q_m + Q_m + \frac{\partial}{\partial t}\left(\frac{\eta S_m}{B_m}\right) = 0 \qquad (13)$$

$$q_m = -\frac{k_{rm}}{B_m \mu_m} k\left(\nabla P_m - \gamma_m \nabla d\right) \qquad (14)$$

where $\nabla \cdot$ is divergence operator; q is volumetric flux or flow rate; η is porosity; Q is a source or sink term, which for the gas phase represents mass transport between the secondary and primary porosity systems; S is fluid saturation; B is the formation volume factor; k is the absolute permeability; k_r is the elative permeability factor; P is fluid pore pressure; γ is fluid unit weight; t is time; μ is viscosity; d is the vertical distance from a given datum; and m refers to each of the fluid phases.

In this formulation, the pore volume is assumed to be fully occupied by the combination of the two fluids, i.e.:

$$S_w + S_{nw} = 1 \qquad (15)$$

where w represents the wetting phase and nw represents the nonwetting phase. The wetting phase and nonwetting phase fluid pressures are assumed to be related as follows:

$$P_{nw} - P_w = P_c \qquad (16)$$

where P_c is the capillary pressure.

COSFLOW can simulate either two-phase or single-phase flow. In the case of a single-phase flow, the pore volume is allowed to be partially filled by the wetting phase fluid, in which case the fluid pore pressure is expressed as:

$$P_w = -P_c \qquad (17)$$

The gas absorbed in the coal seams can enter into the coal cleats as free gas and acts as a source term shown in Equation 13. This process can be described using an approach by Smith and Williams [1984] and Kolesar and Ertekin [1986]. In such an approach, the volume of the adsorbed gas in the coal matrix can be described by the Langmuir adsorption isotherms:

$$V_E = \frac{V_L P_g}{P_L + P_g} \qquad (18)$$

where V_E is the volume of gas that can be adsorbed at pressure P_g (also called the equilibrium concentration); V_L is Langmuir volume, which is the maximum volume of gas that can be adsorbed; and P_L is the pressure at which the volume of the adsorbed gas is half V_L.

The mass transport is described by Equation 19, which is derived from Fick's law:

$$Q_g = \frac{1}{\tau}\left(V_{mat} - V_E\right) \qquad (19)$$

where Q_g is the amount of gas desorbed from the matrix, V_{mat} is the gas volume contained in the matrix, V_E is the matrix gas equilibrium volume, and τ is the so-called sorption time expressed as

$$\tau = \frac{s_f^2}{8\pi D} \qquad (20)$$

where D is the micropore diffusion coefficient, and s_f is coal cleat spacing. This equation includes the proper shape factor for cylindrical matrix elements. Schwerer [1984] has shown that cylindrical matrix elements are adequate for modeling the diffusion process in coals.

Dynamic Coupling

The dynamic interaction between mechanical deformation and fluid flow processes can be described through a set of coupled nonlinear partial differential equations. The presence of a fluid in the mechanical model is considered by using the concept of effective stress so that such a stress field and the pore-fluid pressure satisfy the following force equilibrium conditions:

$$\frac{\partial \sigma'_{ij}}{\partial x_j} - \alpha \frac{\partial P}{\partial x_i} = F_i \qquad (21)$$

Here, σ is effective stress, α is the Biot coefficient, P is pore pressure, F is body force density, x is a spatial coordinate, and i and j indicate the components of the vector and tensor variables in Cartesian space.

The incremental stress changes are related to changes in incremental strain and pore pressure either through linear (elasticity) stiffness terms prior to yielding or through nonlinear (plasticity) stiffness terms after yielding.

Similarly, change in pore volume is used to compute the associated changes in fluid pressures and saturations by solving the following sets of equations:

$$\frac{\eta_0 S_m^0}{B_m^0} = \frac{\eta_1 S_m^1}{B_m^1} \qquad (22)$$

where *0* and *1* refer to initial and final conditions.

The flow of either phase of fluid is controlled by the permeability of the porous medium, which is either derived by field measurements or through theoretical/empirical formulations. There are different formulas proposed in the literature for estimating the permeability of porous medium depending on whether the porous medium is intact or contains a network of fractures. The permeability of a porous rock remains a highly nonlinear dynamic function of mining-induced stress and subsequent fractures. Thus, it is important not only to estimate the initial permeability correctly, but also to compute its possible variation induced by mining.

Kozeny and Ber [1927], Hubbert [1940], Krumbein and Monk [1943], and De Wiest [1969] attempted to establish a relationship between stress and permeability through a definition of hydraulic radius, which is a function of grain diameter, porosity, grain shape, and packing. As rock masses usually contain natural fractures that predominantly control the fluid movements, there is a distinctive advantage in formulating a model on the basis of equivalent fracture network. In that framework, fluid flow through a single fracture can be expressed using a flow-through parallel plate analog where a fracture is idealized as a planar opening with a constant aperture [Bai and Elsworth 1994; Esterhuizen and Karacan 2005].

Such an approach to describe the rock mass permeability through equivalent fracture idealization is well suited for coal measure rocks (i.e., rock masses in a coal mining environment). Seedsman [1996] discussed hydrogeological aspects of Australian longwalls and pointed out that water flow in coal measure rocks in New South Wales and Queensland is dominated by defects, cleats, joints, bedding, and faults rather than via pores in the rock mass.

For a laminar flow within the fracture network, the hydraulic conductivity of a set of parallel fractures with a spacing, *s*, and aperture, *a*, is given by (e.g., Louis [1969]):

$$k = \frac{\rho g a^3}{12 s \mu} \qquad (23)$$

where ρ is the fluid density, g is the gravitational acceleration, and μ is the dynamic viscosity of the fluid.

By assuming a fractured rock mass consists of many interconnected fractures, it may be further idealized as an equivalent porous continuum where the rock mass is represented by an equivalent anisotropic hydraulic conductivity matrix defined in terms of mean fracture spacing and mean aperture. Here it is assumed that the principal directions of this matrix are aligned with the coordinate axes. For a fractured rock with fracture spacing Fs_i (i = 1, 2, 3) and fracture apertures Fa_i (i = 1, 2, 3), the relationships between the absolute initial (premining) permeability components (k_{11}, k_{22}, and k_{33}) and the fracture parameters can be expressed as:

$$k_{11}^{ini} = \frac{Fa_2^3}{12Fs_2} + \frac{Fa_3^3}{12Fs_3} \qquad (24)$$

$$k_{22}^{ini} = \frac{Fa_3^3}{12Fs_3} + \frac{Fa_1^3}{12Fs_1} \qquad (25)$$

$$k_{33}^{ini} = \frac{Fa_1^3}{12Fs_1} + \frac{Fa_2^3}{12Fs_2} \qquad (26)$$

This formulation is amenable to easy evaluation of modifications to the hydraulic conductivities as a function of stress-induced changes in fracture aperture. In this study, change in rock mass permeability is formulated on the basis of the mine-induced strain [Elsworth 1989; Bai and Elsworth 1994; Liu and Elsworth 1997], as follows:

$$k_{11} = \frac{1}{2}k_{11}^{ini}\left[(1+\beta_2\Delta\varepsilon_{22})^3 + (1+\beta_3\Delta\varepsilon_{33})^3\right] \qquad (27)$$

$$k_{22} = \frac{1}{2}k_{22}^{ini}\left[(1+\beta_1\Delta\varepsilon_{11})^3 + (1+\beta_3\Delta\varepsilon_{33})^3\right] \qquad (28)$$

$$k_{33} = \frac{1}{2}k_{33}^{ini}\left[(1+\beta_1\Delta\varepsilon_{11})^3 + (1+\beta_2\Delta\varepsilon_{22})^3\right] \qquad (29)$$

where $\Delta\varepsilon_{ii}$ are the normal strain components and β_i are expressed as:

$$\beta_i = 1 + \frac{1 - R_m}{\left(\frac{Fa_i}{Fs_i}\right)^n} \quad \mathbf{i=1,2,3} \qquad (30)$$

Here, R_m is the modulus reduction ratio (ratio of rock mass modulus to rock matrix modulus), the term Fa_i/Fs_i may be defined as a function of equivalent fracture porosity, and n is a constant (in Liu and Elsworth [1997], n is assumed to be equal to 1.0). Both R_m and n are considered to be fitting parameters and thus need to be calibrated properly against well-documented field data. If R_m equals 1.0, then β_i equals 1.0, resulting in minimal strain-induced permeability changes. When R_m tends to 0.0 (i.e., the case of highly fractured rock), β_i will attain the maximum value and thus will induce large changes in permeability.

MINE SUBSIDENCE PREDICTION

Empirical Methods

Prediction methods can be classified as empirical methods, methods employing influence functions, and methods employing theoretical models [Helmut 1983]. The empirical methods are derived from statistical analyses of previously monitored data and site observation and thus are only applicable to the area where the data have been analyzed. Nevertheless, because of their simplicity, the empirical methods have been widely used. In Australia, two empirical methods are widely used to predict subsidence induced by longwall mining. Both methods are based on extensive subsidence measurement data and observations across a number of coalfields in New South Wales.

The first, the DMR method, was developed by the Department of Mineral Resources, New South Wales, Australia [Holla and Barclay 2000]. Its first version was proposed in 1985 and published in a subsidence handbook for the Southern Coalfield. It was then used for longwall mining planning in the Southern, Newcastle, and Western Coalfields, New South Wales. This method provides several graphs for determining maximum subsidence for single and multiple panels. The strain, tilt, and curvature can also be determined by using the simple formulas provided.

The second, the Incremental Profile Method, was developed by Waddington and Kay [1995]. Its first version was proposed as a subsidence prediction method for Appin and Tower Collieries in the Southern Coalfield in 1994. It was then extended to other collieries of the Southern and Newcastle Coalfields. Unlike the DMR method in which maximum subsidence parameters are predicted, the Incremental Profile Method predicts the incremental subsidence profile for each longwall panel in a series of longwall panels. The respective incremental profiles are then added to form the cumulative subsidence profile at any stage in the development.

Both empirical methods require geometrical parameters as input such as panel width, pillar width, seam height, and mining depth. The effects of any variations in overburden geology, strata properties, and stress conditions cannot be assessed using these methods. A recent study [Strata Engineering Australia 2003] indicated that the geological conditions above the panels could affect the subsidence significantly. Only numerical methods are likely to be able to predict the effect of overburden geology, strata properties, initial stress, or surface topography on subsidence.

COSFLOW: A Numerical Approach

Subsidence due to underground coal mining is a complex process that involves caving, fracturing, and bending of stratified overburden. Such overburden is layered and exhibits highly anisotropic strength and deformation characteristics. This makes it necessary to include effects of stratification into the mathematical formulations describing the load deformation behavior. The most efficient way to study the load deformation behavior of the overburden is to devise an equivalent continuum model where the effect of stratification is incorporated implicitly in the model formulation, e.g., Cosserat model.

Subsidence Simulation: Appin Colliery

Appin Colliery is located in the Southern Coalfield of the Sydney Basin. The longwall mining method is used to extract coal from the Bulli Seam.

The longwall panels modeled at Appin are 200 m wide and the extraction height is 2.3 m in the Bulli Seam, which is at a depth of about 500 m. COSFLOW was used to simulate the overburden movement caused by mining longwall panels 21B to 28A, which were completed from 1991 to 1996. The monitored subsidence data and COSFLOW prediction are shown in the top and bottom of Figure 7, respectively. A comparison indicates that the COSFLOW estimates are generally consistent with the monitored data. It can be seen that COSFLOW successfully simulated subsidence development in detail as mining progressed from panel 21B to 28A. The results also show the effects of chain pillars on the subsidence.

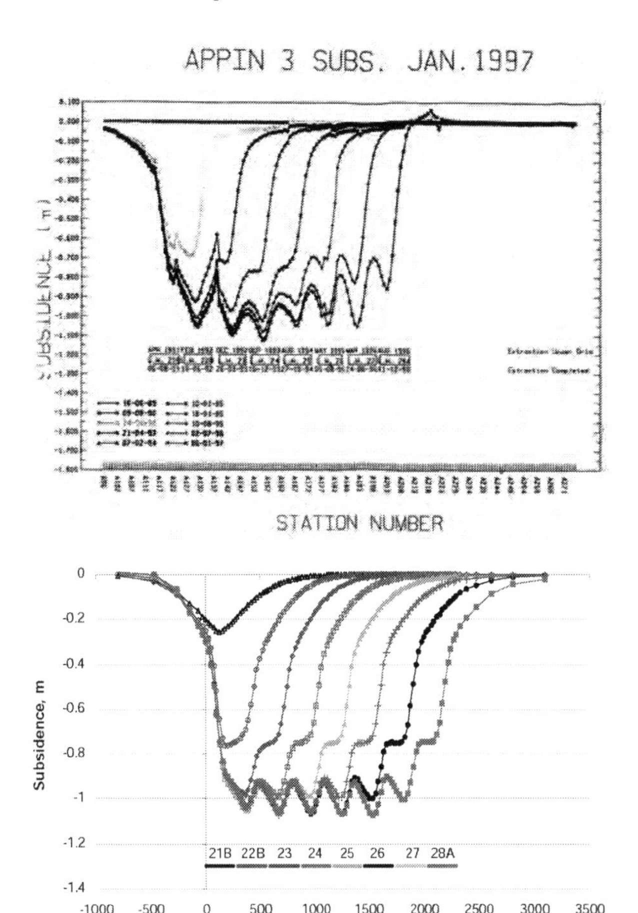

Figure 7.—A comparison of monitored subsidence data and COSFLOW prediction (after Guo et al. [2004]).

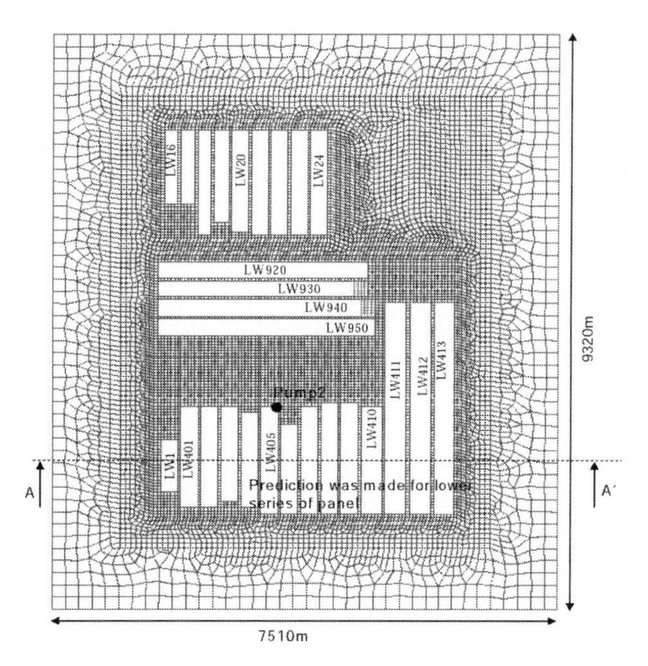

Figure 8.—A plan view of the numerical grid used in the simulation.

MINE GROUNDWATER FLOW SIMULATION: SPRINGVALE MINE

Figure 8 shows the plan view of the finite-element mesh. The location of pump 2 used for dewatering from LW1 to LW405 is also shown. The numerical model consists of about 500,000 finite elements and simulated a region 7.5 km long in the east-west direction and 9.3 km long in the north-south direction, thus covering an area of approximately 71 km^2. The permeability and geomechanical values used in the simulation are presented in Tables 1 and 2.

Table 1.—Permeability values used in the simulation of water inflow

Rock unit	Horizontal hydraulic conductivity (m/sec) × 10^{-8}	Vertical hydraulic conductivity (m/sec) × 10^{-8}	Porosity
Floor	1.0	0.25	0.15
Mining seam	10.0	10.0	0.10
Unit 1	2.5	1.0	0.15
Unit 2	0.05	0.05	0.10
Unit 3	1.0	0.25	0.15
Unit 4	0.1	0.1	0.10
Unit 5	1.0	0.25	0.15

All of the rock units were assumed to dip parallel to the mining seam, which dips at 1° to the east and 0.7° to the north. The actual overburden sequence has been simplified combining lithological layers to represent rock characteristics of primary importance to obtain an average response and a reasonable fit with measurements (see Figure 5). Piezometers installed in the AQ4 horizon seemed to be unaffected by the mining underneath; thus, for simplicity, AQ4 and AQ5 were not included in the numerical simulation.

Figure 9 shows the numerical predictions of water inflow into the lower series of panels (LW6 to LW10). The water inflow rate into the mine up to LW6 was much smaller, with total flow averaging around 55 L/s. However, the measured water inflow rate after mining LW7 onward increased substantially, yielding a rate of about 88 L/s after mining LW7. In general, COSFLOW can be seen to provide accurate predictions of mine water inflow when compared with the mine data. After 3 years, the predicted inflow rates are found to be in good agreement with the measurements.

Table 2.—Geomechanical parameters used in the numerical simulation of water inflow

Rock units	Young's modulus (GPa)	Cohesion (MPa)	Friction angle (degrees)	Tensile strength (MPa)	Remarks
Base	18.0	8.0	35.0	3.7	No bedding
Floor	10.0	2.5	35.0	0.96	No bedding
Mining seam	3.5	1.3	40.0	0.5	No bedding
Unit 1	10.0	3.0	35.0	1.15	0.5-m bed spacing, joint cohesion = 0.5 MPa, joint friction angle = 25°
Unit 2	10.0	3.0	30.0	1.0	0.25-m bed spacing, joint cohesion = 0.3 MPa, joint friction angle = 25°
Unit 3	10.0	3.0	35.0	1.15	0.5-m bed spacing, joint cohesion = 0.5 MPa, joint friction angle = 35°
Unit 4	10.0	3.0	30.0	1.0	0.25-m bed spacing, joint cohesion = 0.3 MPa, joint friction angle = 25°
Unit 5	10.0	3.0	35.0	1.15	0.5-m bed spacing, joint cohesion = 0.5 MPa, joint friction angle = 25°
Top	5.0	2.0	35.0	0.77	No bedding

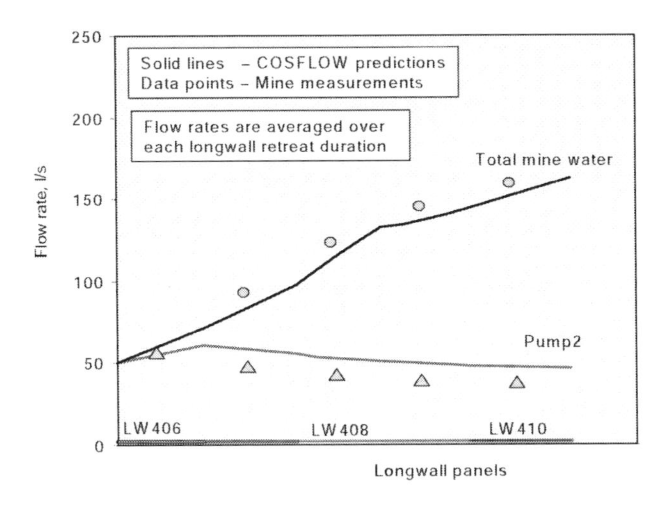

Figure 9.—Comparison of numerical prediction with the mine measurements (after Guo et al. [2006]).

Panels up to LW403 are 255 m wide, panels LW404 to LW409 are 265 m wide, and panels LW410 onward are 315 m wide. The noticeable change observed in the water inflow rates could be attributed to the change in the width of the panel as well as to the so-called multipanel effect. Figure 10 shows the distribution of vertical permeability changes as predicted by the numerical model. It can be clearly seen in the figure that pattern of change in vertical permeability differs for panels up to LW406 and those LW407 onward, indicating a likelihood of increased water inflow.

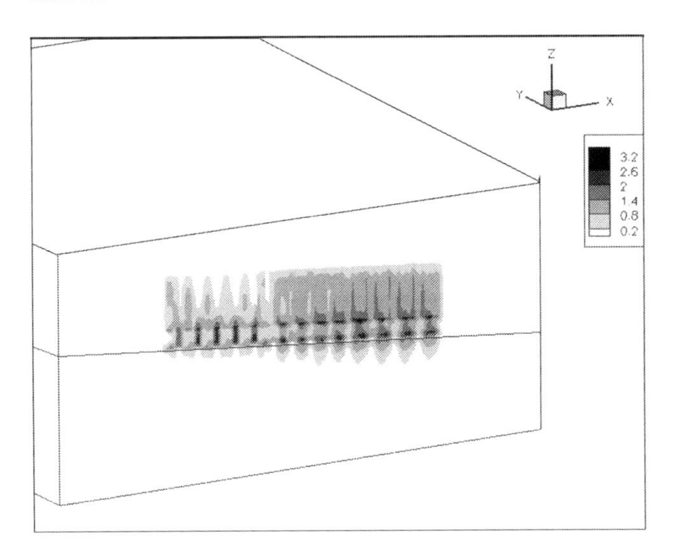

Figure 10.—Change in vertical permeability along the vertical section A–A' in Figure 8 (values indicate orders of magnitude change).

Figure 11 shows the pore water pressure distribution in the coal seam after mining panel LW408 as obtained from the numerical simulation. The location of the piezometer installed in borehole SPR31 is marked on the plot. The pore water pressure at that location is about 400,000 Pa (i.e., water head is about 41 m); the piezometer readings fluctuated around 43 m after LW408 was completed.

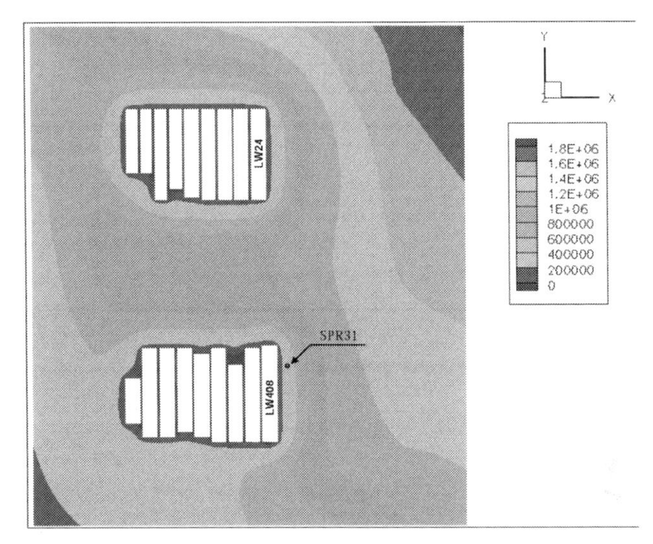

Figure 11.—Pore pressure distribution (in pascals) at the mining seam after mining LW408; note the position of the piezometer, indicated as SPR31 (after Guo et al. [2006]).

METHANE EMISSION SIMULATION: MINE A

The numerical simulation of methane emissions from multiple seams into a longwall panel at an Australian mine is presented in this section. The extraction of two adjacent panels is modeled. The panels are denoted A and B (see Figure 12), where panel A is excavated before panel B. The emphasis on gas emission estimates is for panel B; panel A is included in the simulation simply to provide accurate initial conditions for the mining of panel B. Panel B could represent any longwall panel with similar predrainage and postdrainage conditions and extraction rates in a similar geological regime. The predictions from numerical simulations should be equally valid for comparison with measurements obtained from any similar longwall panels; thus, the numerical results are compared with mine measurements from two longwall panels.

Table 3.—Geomechanical parameters used in the numerical simulation of gas emission

Rock units	Young's modulus (GPa)	Cohesion (MPa)	Friction angle (degrees)	Tensile strength (MPa)	Remarks
Base	10.0	4.0	30.0	1.4	No bedding
Base 1	4.0	2.0	30.0	0.7	0.5-m bed spacing, joint cohesion = 0.8 MPa, joint friction angle = 30°
Base 2	3.0	1.15	30.0	0.4	0.1-m bed spacing, joint cohesion = 0.2 MPa, joint friction angle = 20°
Coal seams	3.5	1.3	40.0	0.5	No bedding
Unit 1a	12.0	3.12	35.0	1.2	No bedding
Unit 2	8.0	2.3	30.0	0.52	0.25-m bed spacing, joint cohesion = 0.5 MPa, joint friction angle = 30°
Unit 3	5.0	1.7	30.0	0.56	0.25-m bed spacing, joint cohesion = 0.05 MPa, joint friction angle = 20°
Unit 4	7.0	2.5	35.0	0.96	0.25-m bed spacing, joint cohesion = 2.0 MPa, joint friction angle = 20°
Unit 5	6.0	1.6	32.5	0.58	0.25-m bed spacing, joint cohesion = 0.1 MPa, joint friction angle = 30°
Top 1	10.0	3.9	35.0	1.5	0.5-m bed spacing, joint cohesion = 1.0 MPa, joint friction angle = 20°
Top 2	7.0	2.08	35.0	0.8	No bedding

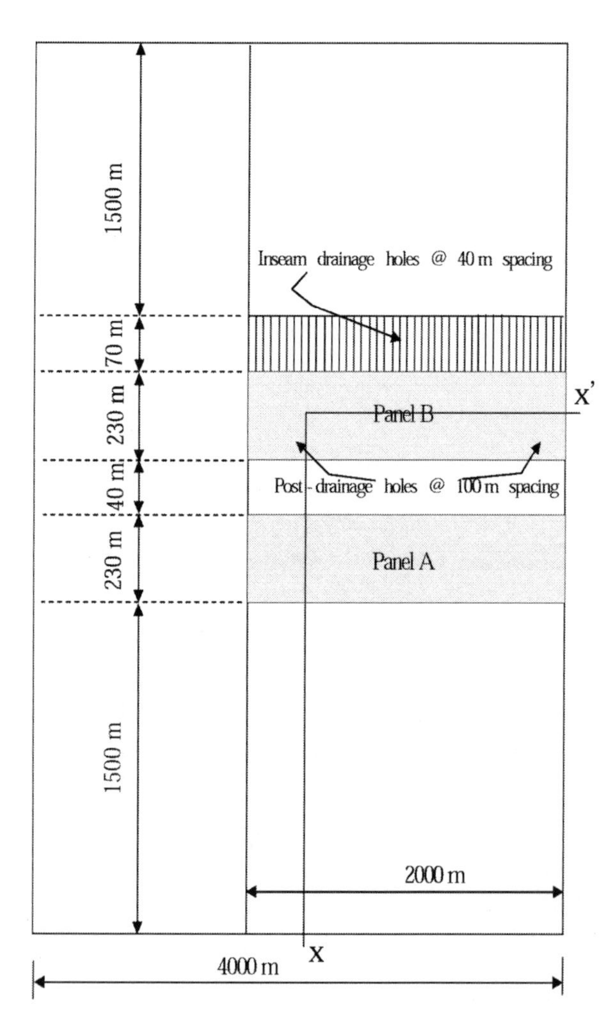

Figure 12.—Simulation layout of mine A.

The mining panels are 230 m wide, 2.6 m high, and about 390 m deep. A total of 260,000 finite elements were used in the simulation. Figure 13 shows the simplified geology used in the simulation. The mining seam and two other seams were included in the simulation. The geomechanical properties adopted for the numerical simulation for the various rock layers are listed in Tables 3 and 4. On the basis of coal seam gas pressure, the water table is assumed to be 306 m above the mining seam, yielding a gas pressure of 3 MPa at the mining seam.

Table 4.—Flow parameters used in the numerical simulation of gas emissions

Parameter	Value
Rock permeability in horizontal direction (md)	30.0
Rock permeability in vertical direction (md)	3.0
Coal permeability in horizontal direction (md) – k_h	6.0–9.0
Coal permeability in vertical direction (md) – k_v	0.6–0.9
Gas content (m^3/t)	13.5
Langmuir volume (m^3/t)	23.8
Langmuir pressure (MPa)	1.5
Coal sorption time (days)	10
Reservoir pressure (MPa)	3.0

In this case, the model used measurements of gas production from predrainage boreholes for calibration. The simulated mining seam is first predrained, using in-seam boreholes (with 96-mm diameter) shown in Figure 12, for a simulated time equal to the actual predrainage time. This allowed calibration of the input parameters and provided initial conditions for the water/gas state before the longwall extraction.

Following the predrainage simulation, panel A was extracted in 10 large steps of 200 m, then 700 m of panel B extraction was simulated in steps of 40 m. As the simulated extraction progressed, vertical boreholes of

254 mm in diameter were added to the model on the tailgate (panel A) side of panel B at 100-m spacing, as shown in Figure 12. Gas may flow into these boreholes or into the longwall panels. These flows are recorded separately for comparison with postdrainage measurements and flow into the ventilation system, respectively.

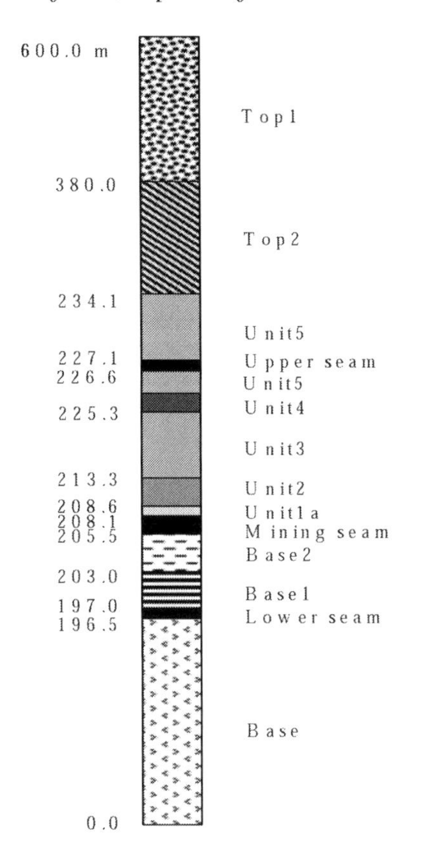

Figure 13.—A geological log used in the simulation of mine A.

Figure 14 compares numerical predictions and actual measurements in the mine. In the figure, methane emission into the ventilation air in one of the longwall panels in the mine is compared with the COSFLOW prediction for panel B. It can be seen that COSFLOW provides accurate predictions of average gas emissions into the longwall panel. Many of the fluctuations in measurements seen in Figure 14 are probably caused by variations in mining extraction rate, interruption to mining and postdrainage operations, variability of local geology and gas content, or effectiveness of predrainage schemes; none of these factors are included in the simulation. In general, COSFLOW can be seen to provide accurate predictions of average gas productions when compared with the mine data.

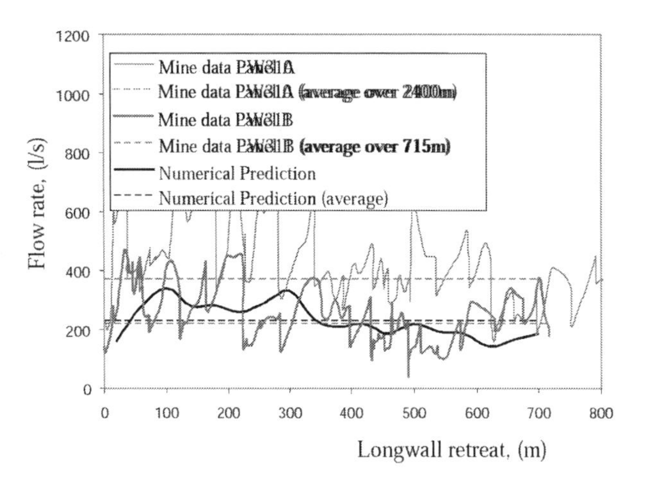

Figure 14.—Comparison of numerical results with the mine measurements of gas emission into a longwall panel.

CONCLUSIONS

Reliable prediction of mining impact on surface subsidence, subsurface aquifers, water inflow, and gas emission into underground coal mines is essential for improving mine safety and reducing coal production costs, as well as for assessing the environmental impact of mining.

This paper describes an integrated approach in assessing the impact of mining, including a new 3-D coupled mechanical two-phase double-porosity finite-element code called COSFLOW developed by CSIRO Exploration and Mining to serve the needs of the mining industry. A unique feature of COSFLOW is the incorporation of Cosserat continuum theory in its formulation. In the Cosserat model, interlayer interfaces (joints, bedding planes) are considered to be smeared across the mass, i.e., the effects of interfaces are incorporated implicitly in the choice of stress-strain model formulation. An important feature of the Cosserat model is that it incorporates bending rigidity of individual layers in its formulation, which makes it different from other conventional implicit models.

For the past 6 years, CSIRO Exploration and Mining has been actively involved in mine site geological, geotechnical, and hydrogeological characterization and predictive simulation of mine subsidence, mine water inflow, and gas emission. In this work, COSFLOW has been found to be capable of producing accurate predictions. The example of mining-induced surface subsidence, mine groundwater inflow, and mine gas emission predictions presented in this paper shows the remarkable capability of COSFLOW to simulate the mining-induced rock deformation, aquifer interference, permeability changes, and water/gas flow.

It is important to note that the predictions made using the integrated approach involving field instrumentation, field monitoring, and relevant data acquisition, and ultimate prediction simulation, can only be as good as the geotechnical and hydrogeological input data. Thus, field data gathering is a critical task.

ACKNOWLEDGMENTS

We would like to express our sincere gratitude to NEDO and JCOAL of Japan for providing a significant part of the funds during the development of COSFLOW. We would also like to thank Springvale Colliery and BHPB Illawara Coal for providing us with the mine data.

REFERENCES

Adhikary DP, Dyskin AV [1997]. A Cosserat continuum model for layered media. Computers and Geotechnics *20*(1):15–45.

Adhikary DP, Dyskin AV [1998]. A continuum model of layered rock masses with non-associative joint plasticity. Int J Numer Anal Meth Geomech *22*(4):245–261.

Adhikary DP, Guo H [2002]. An orthotropic Cosserat elasto-plastic model for layered rocks. Rock Mech Rock Eng *35*(3):161–170.

Adhikary DP, Guo H, Chen S [2002]. LW410 subsidence estimate at Springvale colliery. CSIRO Exploration and Mining Report 1221C, 17 pp.

Adhikary DP, Guo H, Shen B, Knight A [2004]. Interpretation of hydrogeological data at Springvale colliery. Exploration and Mining Report P2004/62, 28 pp.

Alehossein H, Carter JP [1990]. On the implicit and explicit inclusion of joints in the analysis of rock masses. In: Rossmanith H-P, ed. Mechanics of jointed and faulted rocks. Rotterdam: Balkema, pp. 487–494.

Bai M, Elsworth D [1994]. Modelling of subsidence and stress-dependent hydraulic conductivity for intact and fractured porous media. Rock Mech Rock Eng *27*(4):209–234.

Cosserat E, Cosserat F [1909]. Théorie des corps déformables (in French). Paris: Hermann.

De Wiest RJM, ed. [1969]. Flow through porous media. London: Academic Press.

Elsworth D [1989]. Thermal permeability enhancement of blocky rocks: one dimensional flows. Int J Rock Mech Min Sci Geomech Abstr *26*(3/4):329–339.

Esterhuizen GS, Karacan CÖ [2005]. Development of numerical models to investigate permeability changes and gas emissions around longwall mining panels. In: Chen G, Huang S, Zhou W, Tinucci J, eds. Proceedings of the 40th U.S. Rock Mechanics Symposium (Anchorage, AK, June 25–29, 2005). Alexandria, VA: American Rock Mechanics Association, pp. 1–13.

Gerrard CM [1982]. Elastic models of rock masses having one, two and three sets of joints. Int J Rock Mech Min Sci and Geomech Abstr *19*:15–23.

Gerrard CM, Pande GN [1985]. Numerical modelling of reinforced jointed rock masses. I. Theory, computer and geotechnics. Vol. 1, pp. 293–318.

Guo H, Shen B, Poulsen B [2002]. Review of existing ground water and geotechnical data at Springvale colliery. CSIRO Exploration and Mining Report, 20 pp.

Guo H, Adhikary DP, Xue S, Craig MS, Poulsen BA, Chen S, Wendt M, Su S, Mallett C [2003a]. Pre-development studies for mine methane management and utilisation. CSIRO Exploration and Mining Report 1079C, 134 pp.

Guo H, Shen B, Poulsen B, Zhou B [2003b]. Interpretation of geological, hydrogeological and ground water data at Springvale colliery. Exploration and Mining Report 1129C, 90 pp.

Guo H, Chen SG, Craig MS, Adhikary DP, Poole G [2004]. Coal mine subsidence simulation using COSFLOW. AusIMM J *Mar/Apr*(2):47–52.

Guo H, Adhikary DP, Rutzou P, Miller B [2006]. Integrated approach to mine ground water assessment. In: Proceedings of Water in Mining (November 14–16, 2006).

Helmut K [1983]. Mining subsidence engineering. Berlin, Germany: Springer-Verlag, pp. 183–249.

Holla L, Barclay E [2000]. Mine subsidence in the southern coalfield, NSW, Australia. New South Wales, Australia: Department of Mineral Resources.

Hubbert MK [1940]. The theory of ground water motion. J Geol *48*:785–944.

King GR, Ertekin TM [1989a]. A survey of mathematical models related to methane production from coal seams. Part I: Empirical and equilibrium sorption models. In: Proceedings of the International Coalbed Methane Symposium, pp. 125–138.

King GR, Ertekin TM [1989b]. A survey of mathematical models related to methane production from coal seams. Part II: Non-equilibrium sorption models. In: Proceedings of the International Coalbed Methane Symposium, pp. 139–155.

King GR, Ertekin TM [1991]. State-of-the-art modelling for unconventional gas recovery. SPE Formation Evaluation, pp. 63–71.

King GR, Ertekin TM [1994]. A survey of mathematical models related to methane production for coal seams. Part III: Recent developments (1989–1994). In: Proceedings of the International Coalbed Methane Extraction Conference, pp. 124–128.

King GR, Ertekin TM [1995]. State-of-the-art modelling for unconventional gas recovery. Part II: Recent developments (1989–1994). SPE paper 29575. Rocky Mountain Region Low Permeability Reservoir Symposium and Exhibition, pp. 173–191.

Knight A, Miller B [2005]. Water management in longwall operations. Longwall 2005, Hunter Valley, New South Wales, Australia, October 17–19, 2005.

Kolesar J, Ertekin TM [1986]. The unsteady-state nature of sorption and diffusion phenomena in the micropore structure of coal. SPE Unconventional Gas Technology Symposium, pp. 289–314.

Kozeny J, Ber S [1927]. Wiener Akad, Abt. 2a, 136,271.

Krumbein WC, Monk GD [1943]. Permeability as a function of the size parameters of unconsolidated sand. Trans Am Inst Min Metall Petrol Eng *151*:153–163.

Langmuir I [1916]. The constitution and fundamental properties of solids and liquids J Am Chem Soc *38*:2221–2295.

Lifshitz IM, Rosenzweig LN [1946]. On the theory of elastic properties of polycrystals. J Exp Theor Phys *16*(11).

Lifshitz IM, Rosenzweig LN [1951]. On the theory of elastic polycrystals [letter to the editor]. J Exp Theor Phys *21*(10).

Liu J, Elsworth D [1997]. Three-dimensional effects of hydraulic conductivity enhancement and desaturation around mined panels. Int J Rock Mech Min Sci *34*(8): 1139–1152.

Louis C [1969]. Groundwater flow in rock masses and its influence on stability. Imperial College, U.K.: Rock Mech Res Report 10.

Rutqvist J, Wu Y-S, Tsang C-F, Bodvarsson G [2002]. A modelling approach for analysis of coupled multiphase fluid flow, heat transfer, and deformation in fractured porous rock. Int J Rock Mech Min Sci Geomech Abstr *39*(4):429–442.

Salamon MDG [1968]. Elastic moduli of stratified rock mass. Int J Rock Mech Min Sci Geomech Abstr *5*:519–527.

Schwerer FC [1984]. Development of coal-gas production simulators and mathematical models for well-test strategies. Chicago, IL: Gas Research Institute, final report No. GRI–84/0060.

Seedman R [1996]. A review of the hydrogeological aspects of Australian longwalls. Symposium on Geology in Longwall Mining (November 12–13, 1996), pp. 269–272.

Smith DM, Williams FL [1984]. Diffusional effects in the recovery methane from coalbeds. SPE J *24*:529–535.

Strata Engineering Australia [2003]. Review of industry subsidence data in relation to the influence of overburden lithology on subsidence and an initial assessment of a sub-surface fracturing model for groundwater analysis. Australian Coal Association Research Program (ACARP) report No. C10023.

Waddington AA, Kay DR [1995]. The incremental profile method for prediction of subsidence, tilt, curvature, strain over a series of panels. In: Proceedings of the Third Triennial Conference on Buildings and Structures Subject to Ground Movement. Newcastle, New South Wales, Australia: Mine Subsidence Technological Society, pp. 189–197.

Zienkiewicz OC, Pande GN [1977]. Time dependent multilaminate model of rocks: a numerical study of deformation and failure of rock masses. Int J Numer Anal Meth Geomech *1*:219–247.

A REVIEW OF RECENT EXPERIENCE IN MODELING OF CAVING

By Mark Board[1] and M. E. Pierce[2]

INTRODUCTION

The assessment of the initiation and growth of caving in rock masses has become more important as mining in hard rock and coal have moved toward higher-production, lower-cost methods. In hard rocks, where block, panel, or sublevel caving is used, empirical methods, based on rock mass characterization, combined with experience are typically used to estimate the hydraulic radius for sustained cave growth and the resulting "break" angles and propagation rates of the cave as it grows to the ground surface (e.g., Laubscher [2000]). Currently, there are a large number of caving projects worldwide in the conceptual stages of design, and many of these will be initiated below open pits. The economics of the transition between open-pit and underground mining often is optimized when simultaneous production from the open pit and underground occurs. The shape and timing of the cave propagation, surface breakthrough, and subsidence area are critical to planning of infrastructure location, as well as mining geometry and timing of transition from open-pit mining to underground mining. For this reason, the use of numerical models to simulate the undercutting, draw, cave propagation, and surface subsidence processes, combined with empirical predictions, is becoming more commonplace.

There are numerous numerical modeling programs and approaches available for performing stress and deformation analysis in geomechanics. The important aspect of modeling of caving is not necessarily the numerical program itself, but the methodology for simulating the caving process and the estimation of input material models and properties. This paper describes an approach for estimating rock mass material properties and modeling of caving developed as part of the Mass Mining Technology (MMT) research project (University of Queensland, Australia) and used in simulating cave response at a number of existing and planned caving projects. An algorithm to simulate the caving process was developed within the macro language (FISH) provided with the FLAC3D and 3DEC programs (Itasca Consulting Group, Inc.). An example of application of this caving algorithm to back analysis of cave growth at Palabora Mine in South Africa is presented.

[1]Itasca Consulting Group, Denver, CO.
[2]Itasca Consulting Group, Minneapolis, MN.

CONCEPTUAL MODEL OF THE CAVING PROCESS

The caving process involves undercutting of the ore body by driving a series of parallel tunnels across the ore zone, drilling and blasting the pillars between them on retreat, and pulling the swell from draw points on the production level below. When the hydraulic radius of the undercut has reached a critical dimension, a self-sustained cave will develop as long as the broken and bulked ore is withdrawn. A conceptual model of the developing cave, described by Duplancic and Brady [1999], consists of four main behavioral regions (Figure 1):

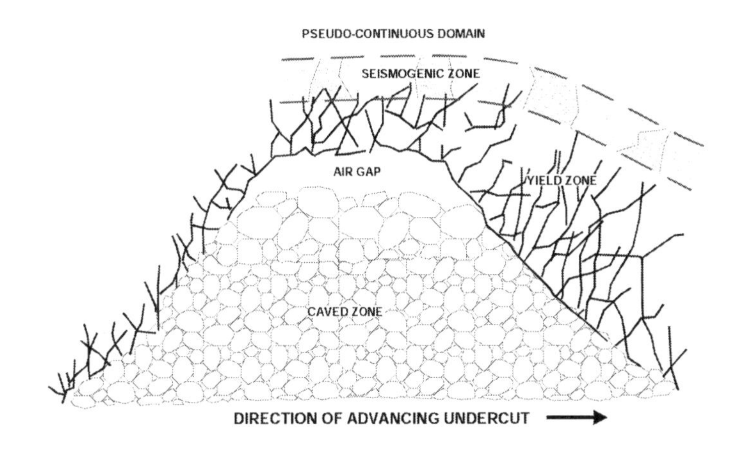

Figure 1.—Conceptual model of caving (after Duplancic and Brady [1999]).

1. ***Pseudocontinuous domain:*** The host rock mass around the caving region behaves mainly elastically. Rock mass behavior and properties are those of an "undisturbed" rock mass.
2. ***Seismogenic zone:*** Microseismic (and sometimes seismic) activity occurs in this region mainly because of discontinuity damage (discontinuities going from peak to residual strength) and the initiation of new fractures.
3. ***Yielded zone:*** The rock mass in this region surrounding the cave is fractured and has lost some or all of its cohesive strength and provides minimal support to the overlying rock mass. Stress components within this region are typically low. This zone is relatively thin in caves that propagate under low-stress conditions ("low" relative to the rock mass strength) and may be very thick in

caves propagating under high-stress conditions ("high" relative to rock mass strength). (This zone was originally called the "zone of discontinuous deformation" by Duplancic and Brady [1999].)

4. ***Caved zone:*** This region consists of rock blocks that have detached from the rock mass and are moving toward the draw points in response to draw and thus might also be called the mobilized zone. The air gap shown in Figure 1 will exist only if the overlying yielded zone retains some level of cohesion.

APPROACH TO REPRESENT CAVING IN A CONTINUUM-BASED NUMERICAL MODEL

Two basic modeling approaches have been used for simulation of caving: discontinuum-based and continuum-based methods. Theoretically, discontinuum-based approaches are preferable for simulating the caving process as one could explicitly account for rock jointing and the stress-related breakage of intact rock blocks, as well as model the movement of the caved rock by several hundred meters. However, the computational requirement to perform such a massive analysis is not possible. Therefore, to make the problem computationally tractable, an algorithm to represent the primary mechanisms of undercutting, draw, and cave propagation within the three-dimensional, continuum-based program FLAC3D was developed. The flowchart in Figure 2 shows the basic features of this algorithm and its implementation using rock mass properties derived empirically and using the Synthetic Rock Mass (SRM) approach. The steps in implementing the caving algorithm are described below.

Geotechnical Database

The estimation of material properties for caving analysis begins with gathering of an extensive geotechnical database. This database is typically gathered from exploration drilling, underground drifting, and open-pit mapping, if available. The most important parameters include:

- Geology
    - Three-dimensional shapes of lithology of the major rock units
    - Geometry, continuity, and surface conditions of major fault structures and shear zones

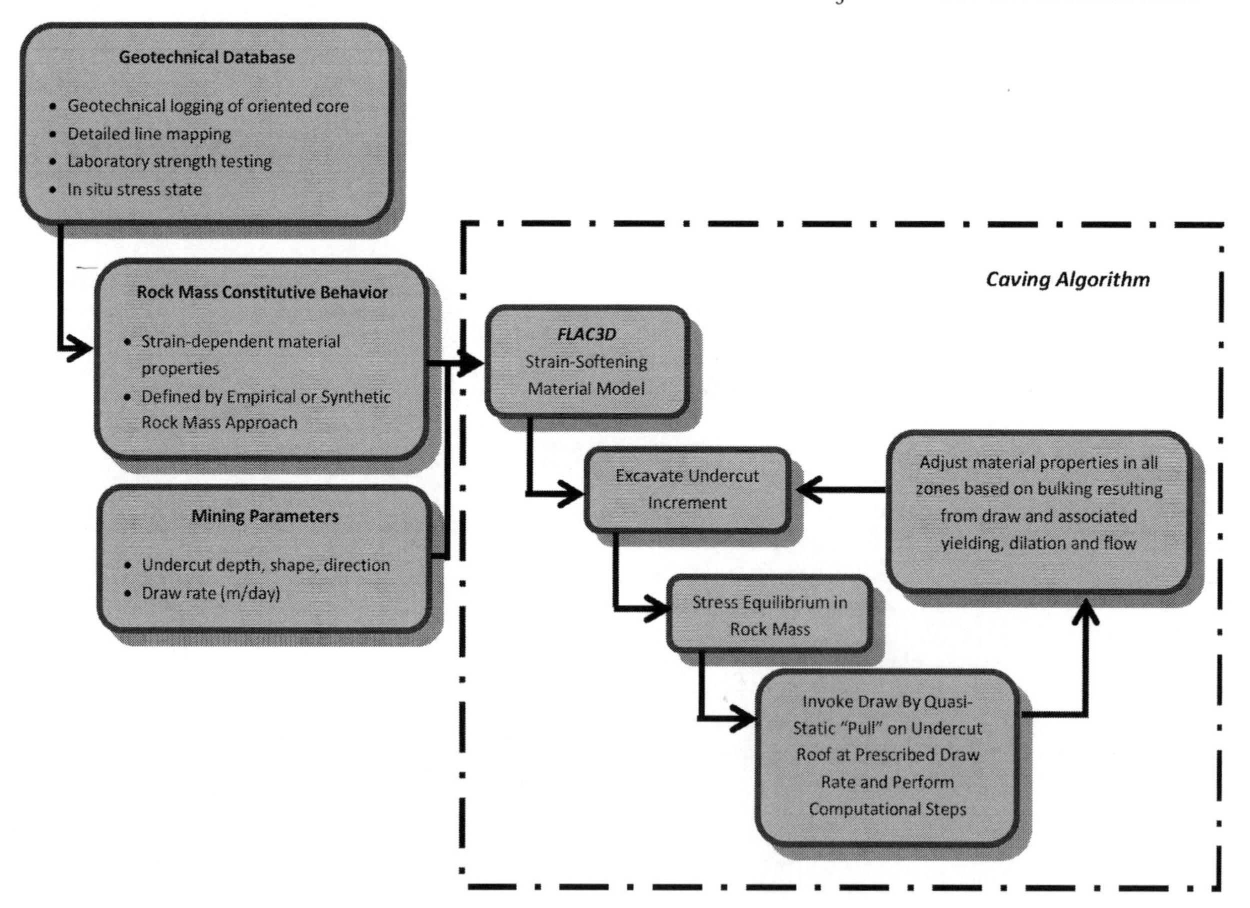

Figure 2.—Flowchart showing caving simulation within the FLAC3D program.

- Geotechnical logging and mapping
  - Oriented core, which provides the orientation, spacing, and surface condition of joint sets
  - Intact rock strength estimate from hammer blows or point load measurements
  - Occurrence of alteration and microdefects, including veinlets, small-scale fracture fabric, etc.
  - Detailed line or pit cell mapping, including measures of fracture orientation, frequency, and trace length
  - Rock mass characterization parameters such as the Geological Strength Index (GSI), Rock Mass Rating (RMR), Mining Rock Mass Rating (MRMR), and Q′.
- Laboratory testing (uniaxial and triaxial compression and tensile strength testing)
- In situ stress state

This information is often developed in the form of a geotechnical model in which geotechnical domains with similar rock mass quality and structure are grouped.

Rock Mass Constitutive Behavior to Represent Caving

Strain-Softening Material Response

A numerical model that represents the caving process must account for the progressive failure and disintegration of the rock mass from an intact/jointed to a caved material. In this complex process, creation of the cave results in (1) deformation and stress redistribution of the rock mass above the undercut; (2) failure of the rock mass in advance of the cave, with associated progressive reduction in strength from peak to residual levels; and (3) dilation, bulking, fragmentation, and mobilization of the caved material. The failure process is characterized by shearing along preexisting joint surfaces and stress-induced fracturing of intact rock blocks. The failure process will necessarily require shear or tensile failure of intact rock bridges between joint segments as the rock mass fragments. This overall process—loading of the rock mass to its peak strength, followed by post-peak reduction in strength to some residual level with increasing strain—is often termed a "strain-softening" process and is the result of strain-dependent material properties.

In the caving model described here, the strain-softening material is described by the Mohr-Coulomb failure criterion in which the post-peak strength behavior is a function of plastic[3] shear-strain-dependent rock mass cohesion (c) and angle of friction (φ).[4] Accumulated plastic shear strain (more specifically, the second invariant of the deviatoric plastic strain tensor) is a common metric for irreversible shear strains in geomaterials and, in a more general sense, can be considered as a measure of damage. Unfortunately, little is known about the relationship between cohesion loss (and friction gain) and plastic shear strain, particularly on a rock mass scale, and there are few guidelines for selection of values for use in modeling. The plastic shear strain required in going from peak strength to a fragmented rock mass (termed here the "critical plastic strain") in the periphery of the cave defines the brittleness of the rock mass failure and may be related to the GSI of the material. This brittleness impacts both the cavability of a given unit and the rate at which a cave will propagate in height for a given amount of draw. Some generalizations may be made regarding these effects. For example, a higher-quality rock mass (higher GSI) with greater solid rock volume participating in the failure process will often act in a more brittle fashion and thus have a lower critical strain value. Conversely, a lower-quality rock mass (lower GSI) with higher fracture frequency will often act in a more ductile fashion and thus have a larger value of the critical strain. An estimate of the relationship between the critical strain and GSI was determined by back analysis of rock mass failure in caves and other openings as a part of the MMT project [Lorig 2000] and provides a starting point for describing the degree of strain softening to be used in simulation of caving. Work is currently under way to more rigorously determine the full stress-strain response of a jointed rock mass (and thus the critical strain) by using the SRM approach, which is described later.

Empirical Approach to Defining Rock Mass Properties

As stated above, the caving algorithm assumes the rock mass strength is represented by a Mohr-Coulomb failure criterion, which is described by the cohesion and angle of friction. One methodology for estimating the mean and range of rock mass strength parameters is the use of the standard Hoek-Brown methodology. Laboratory uniaxial and triaxial testing is used to define the mean and distribution of the uniaxial compressive strength (UCS) and m_i values to determine the intact rock Hoek-Brown envelopes. Geotechnical drill core logging or mapping of rock face exposures is used to estimate the mean and distribution of the rock mass quality (GSI) for each geotechnical domain. Equivalent Mohr-Coulomb strength parameters (c and φ) describing each geotechnical domain are determined from a tangent line to the Hoek-Brown envelope at a low value of confining stress (normally <5 MPa) that reflects the approximate stress conditions near the cave back. An alternative approach, described below, is the use of the SRM methodology to estimate rock mass strength parameters directly.

[3]"Plastic" strain refers to shear strain after peak strength is reached.

[4]A complete description of the Mohr-Coulomb strain-softening material model can be found in the FLAC3D User's Manual (Itasca Consulting Group, Inc.).

The rock mass modulus is calculated from the empirical relation of Serafim and Pereira [1983] as modified by Hoek and Brown [1997]:

$$E_{rockmass}(GPa) = 10^{\frac{GSI-10}{40}} \text{, } UCS_{intact} \geq 100 \text{ MPa} \quad (1)$$

$$E_{rockmass}(GPa) = \sqrt{\frac{UCS_{intact}}{100}} 10^{\frac{GSI-10}{40}} \text{, } UCS_{intact} < 100 \text{ MPa} \quad (2)$$

Impact of Bulking and Dilation

The bulking and dilation of the rock that accompany softening (resulting from creation and opening of new fracture surfaces), as well as the corresponding decrease in density and modulus, must be accounted for to ensure mass conservation (i.e., no mass is created or destroyed during the failure process) and realistic representation of caved rock within the model. The modulus of the rock mass is expected to drop as the rock mass yields, dilates, and bulks. Based on a literature review of caved rock and rockfill properties, Lorig [2000] suggests that the modulus of caved rock at a porosity of 0.3 (a reasonable value for the caved rock) is approximately 250 MPa. This porosity is equivalent to a bulking factor of 0.43, or 43%. Based on these estimates, the modulus of the rock mass is assumed to drop linearly from its in situ value to a value of 250 MPa at a bulking factor of 43% for caved rock.

The dilation angle of the rock mass is assumed to be equal everywhere to 10° based on guidelines provided by Hoek and Brown [1997]. In the caving algorithm, the bulking within each zone (element) of the FLAC3D model is tracked throughout the simulation. Once a zone reaches the user-defined maximum bulking factor, its dilation angle is set to zero. Finally, the density is set for each zone based on lithology. During the simulation of the caving process, the zone density is adjusted automatically to reflect the volumetric changes that accompany bulking according to the following relation:

$$\rho d = \rho s / (1 + B) \quad (3)$$

where ρd = dry density of caved rock;
ρs = solid density of in situ rock;
B = bulking factor = $n/(1 - n)$;
and n = porosity.

Implementation of the Caving Algorithm in FLAC3D

The caving algorithm is implemented in FLAC3D by simulating the undercutting and draw process and the subsequent growth of the yielded and caved zones. Essentially, the model represents the process of undercutting and draw as closely as possible to how they actually occur in the mine. Undercut advance is simulated in the model by converting the model zones (elements) that represent the undercut to fully fragmented and bulked rock. The fragmented and bulked rock are represented by setting their properties to those of broken rock ($c = 0$, $\varphi = 42°$, $E = 250$ MPa) and by setting all stress components to zero. The model is computationally cycled to allow the surrounding rock mass to equilibrate and arrive at a region of yield and redistributed stress around the undercut region. To start the simulation of draw, the undercut volume is deleted and the support to the surrounding rock mass is replaced with equivalent boundary forces that exist *after* the undercut is replaced by the fragmented, bulked rock. Draw of the ore is then simulated by applying a small downward velocity to all grid points in the roof of the undercut. This velocity is set low enough to ensure *pseudostatic* equilibrium throughout the model (i.e., to allow natural gravitational flow of the material and to avoid dynamic "pulling" of the overlying material). The undercut is advanced and draw simulated in many small computational steps at the prescribed draw rate in terms of meters per year.

The mass, m, of material "drawn" from an individual undercut zone during a numerical solution increment is calculated as follows:

$$m = V\, t\, A\, \rho \quad (4)$$

where V = average velocity of the zone nodes in the undercut roof (m/sec);
t = elapsed model time;
A = zone area in plan;
and ρ = zone density.

By summing the masses drawn by all the nodes (grid points), the total production from the cave within the model may be calculated. As the mass is drawn at the undercut, the yielded zone will spontaneously form within the FLAC3D model (dictated by the stress state and yield strength of the rock mass) and *may* progress upward from the undercut. As draw continues, rock within the yielded zone will have moved a sufficient distance (typically >1 m) to be classified as caved (mobilized) material. The extent and growth rate of these two zones (yielded and caved) are both functions of the stress state and material properties of the rock mass (Figure 3). The hydraulic radius at sustained cave growth is a function of the rock mass strength and in situ stress state, the shape of the cave is affected by the rock mass strength and its anisotropy, and "rate" of cave growth (i.e., the height of growth of the yielded zone for an increment of draw) is strongly impacted by the post-peak-strength brittleness of the rock mass.

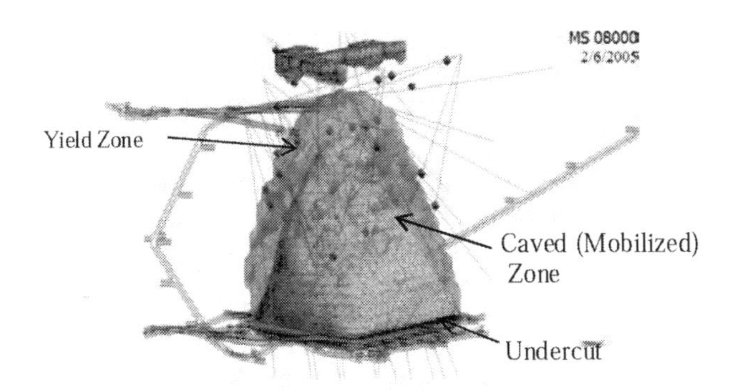

Figure 3.—FLAC3D simulation of Northparkes E26 Lift II cave showing the predicted yielded (cohesion loss) zone surrounding the mobilized cave zone [Pierce et al. 2006].

The Synthetic Rock Mass (SRM) Approach to Defining Rock Mass Constitutive Behavior

Earlier we described the use of empirical methods for defining rock mass material properties for use in the caving model. An alternative and more rigorous approach to defining material properties and overall constitutive response of the rock mass is the SRM method (e.g., Pierce et al. [2007]). The SRM approach provides a platform for simulating the stress-strain response of large-scale "samples" of a rock mass. The basic information typically gathered during geotechnical characterization activities (laboratory uniaxial and triaxial testing of intact rock, direct shear testing of joints, and detailed line mapping of the orientation and continuous length of rock fracturing) is combined to form a "virtual" or "synthetic" rock mass that can be represented within the PFC3D (Itasca Consulting Group, Inc.) numerical modeling approach (Figure 4). The PFC3D model can then be used as a "numerical laboratory" to subject the synthetic samples to stress paths of interest and study the details of failure mode and overall rock mass constitutive behavior.

The first step in the development of an SRM model is to develop an assemblage of bonded spheres in PFC3D representing the intact rock (Figure 4). A representative network of fractures for a given geotechnical domain or lithology is then "mapped" on top of the bonded sphere assembly. The representative fracture geometry is referred to as a discrete fracture network (DFN). The DFN, developed from statistical analysis of field fracture mapping data, is a stochastic representation of the heterogeneous nature of the geometry and continuity of rock mass fracturing for a given lithology or geotechnical domain. A number of programs are available to produce DFNs of disc-shaped fractures from the statistical results of detailed line mapping on tunnel or slope bench faces. Examples include FracMan (Golder Associates, Inc.) and 3FLO (Itasca Consultants S.A.S.). The recent development of a new smooth-joint contact model within the PFC3D [Mas Ivars et al. 2008] makes it possible to realistically simulate the shearing behavior of large numbers of non-persistent joints and the subsequent fracture of intervening solid rock bridges.

The process of exercising an SRM model begins with calibrating the interparticle bonds in PFC3D to reproduce the intact rock stress-strain response (Figure 5). The modulus and peak strength are calibrated against laboratory data for uniaxial and triaxial loading conditions. If the "intact" rock blocks in the rock mass are larger than tested in the laboratory or have extensive microdefects or veining, then this should be reflected in the rock block strength used for calibrating PFC3D.

Following the intact rock calibration, SRM rock mass "specimens" are generated to test the stress-strain characteristics of the jointed rock. One or more DFNs are generated for each lithology at a scale appropriate to represent the rock mass scale (e.g., 10-m cube or larger), and the structure is mapped onto an array of PFC3D particles. These SRM samples may contain upwards of a million

Figure 4.—A synthetic rock mass "sample" *(right)* is created by superimposing a stochastically defined discrete fracture network *(center)* onto a PFC3D particle solid *(left)*.

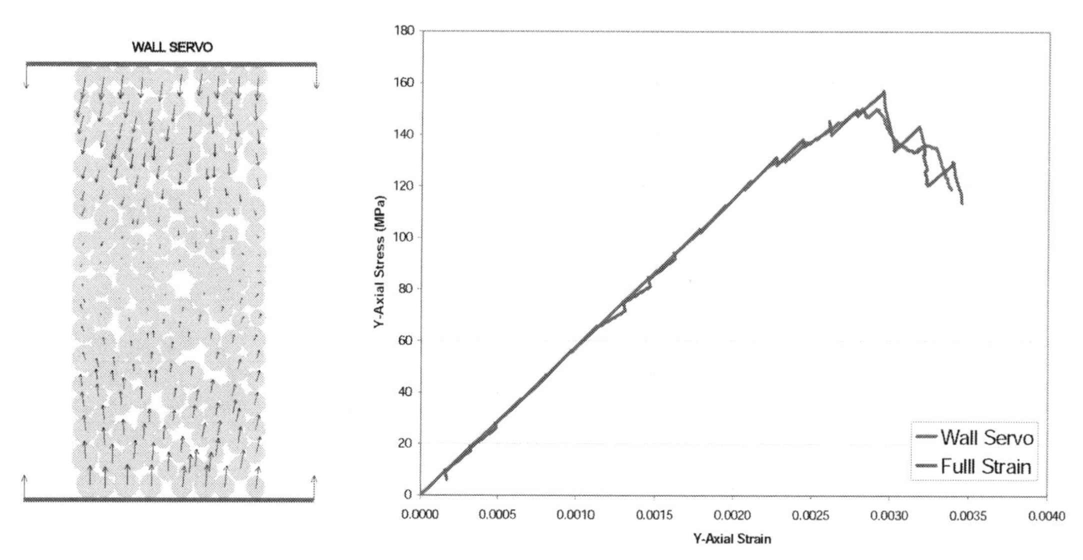

Figure 5.—Uniaxial compression test calibration for intact rock material extracted from an SRM (adapted from Mas Ivars et al. [2008]).

particles and tens to hundreds of thousands of fractures. The SRM samples may be subject to site-specific stress paths or to more standard uniaxial and triaxial compression with driving stresses in principal directions (e.g., vertical, north-south, and east-west) to determine the impact of anisotropy introduced by the joint fabric on rock properties. Failure envelopes are fit to the data via the same procedures used in the laboratory. Size-strength relations can also be used to verify that the scale of the SRM results conform to the scale of the problem to be analyzed.

SRM APPROACH AND CAVING ALGORITHM APPLIED TO BACK ANALYSIS OF CAVE BEHAVIOR AT PALABORA MINE

Introduction

Open-pit operations at Rio Tinto's Palabora copper mine in South Africa were initiated in 1964 and completed in 2000 when the pit reached a depth of approximately 800 m. Panel caving operations were initiated at the completion of the open pit by undercutting of the ore body at a depth approximately 400 m beneath the pit bottom, building to a production rate of 25,000 to 30,000 tpd. In late 2003, crack development was noted in the base of the pit above the cave, which was followed in 2004 by a large slope failure on the north and northwest wall of the pit, extending to the pit crest (Figure 6). The extent of this pit

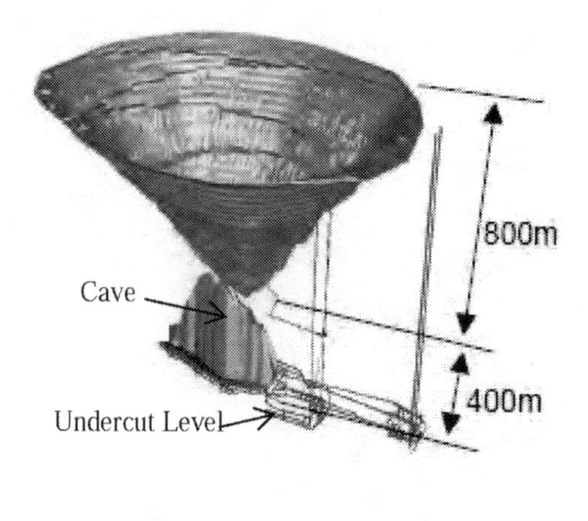

Figure 6.—Slope failure in Palabora pit *(left)* induced by cave from underground mine *(right)*.

wall failure was not expected. Back analysis by Brummer et al. [2006] attributed the failure mechanism to steeply dipping, pervasive joint sets that form wedges that daylight into the cave back below, thus undermining the toe of the north slope

Following is a description of an application of the SRM approach to define the rock mass constitutive behavior, particularly the carbonatite ore body, and to apply these estimates in the FLAC3D caving algorithm for back analysis of the cave and pit wall failure. Presumably, if the anisotropy of the rock mass material behavior is correctly represented by the SRM (with its embedded DFN), the cave propagation and slope failure should be represented in a reasonably accurate prediction.

SRM Analysis of the Rock Mass

The rock mass at Palabora contains four major geotechnical domains for which SRM analyses were conducted. Here, the results of the SRM studies for the primary carbonatite ore host are described. The carbonatite is a hard, strong rock (UCS ~140 MPa) with relatively widely spaced (mean frequency 0.77/m), vertically persistent (trace length mean of 15 m) fracture sets. A DFN for the carbonatite was constructed from the statistical distribution of fracturing defined by underground detail line mapping as well as open-pit bench mapping (Figure 7).

Figure 8 shows an 80-m cube of carbonatite SRM that has been subdivided into samples of successively smaller dimension. These samples were then subjected to a series of compression tests with primary loading directions in the principal (vertical, north-south, east-west) directions. The SRM results clearly show size effect and anisotropy of

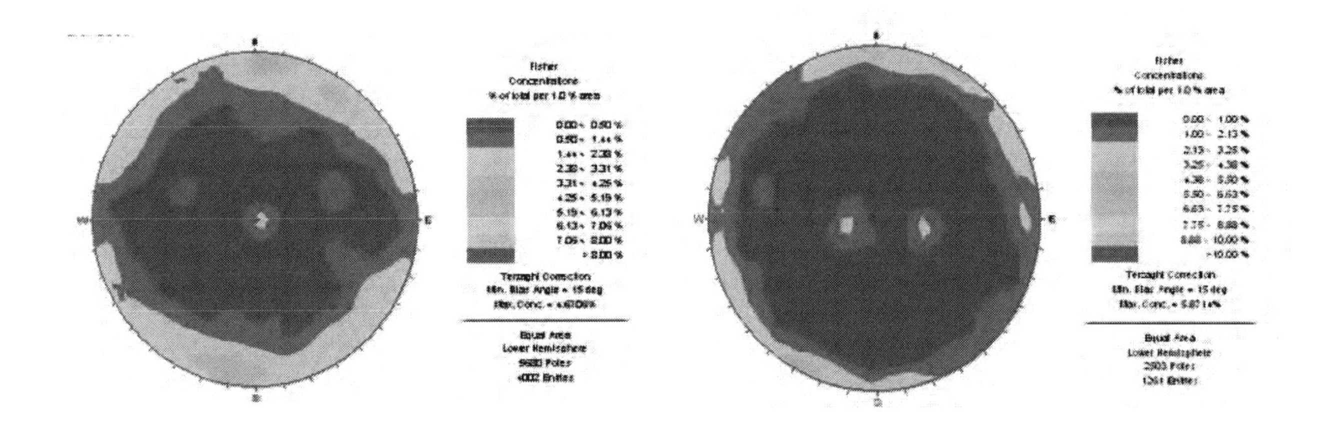

Figure 7.—Vertically oriented joint sets in carbonatite from mapping in the underground line mapping *(left)* and open-pit mapping *(right)* [Mas Ivars et al. 2008].

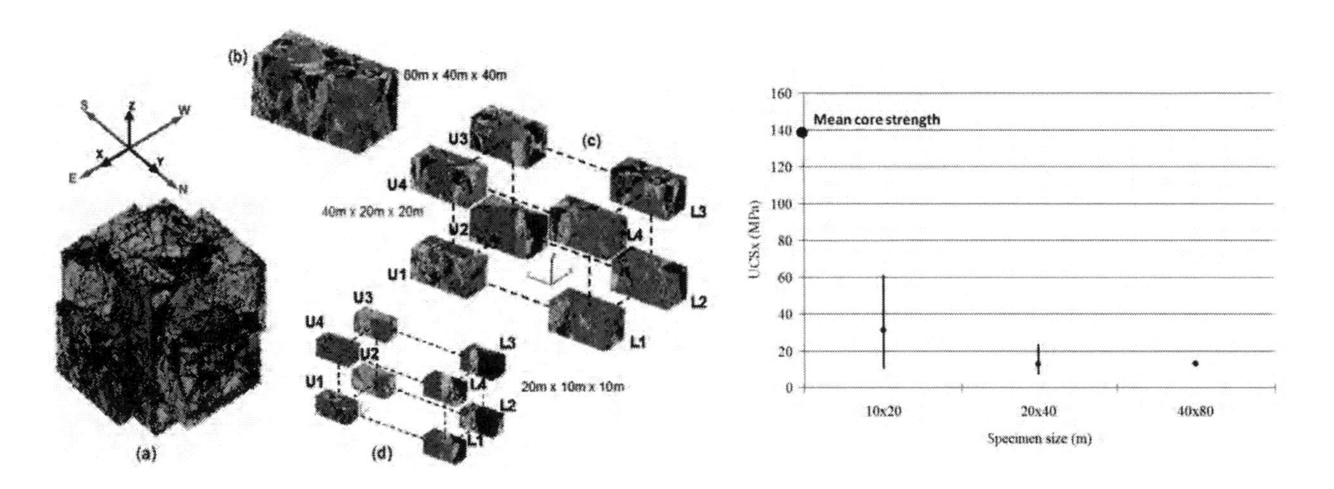

Figure 8.—An 80-m cube SRM jointed sample (a) subdivided into a series of samples of various sizes. The variation in UCS with sample size *(right)* shows the size effect on rock mass strength (adapted from Mas Ivars et al. [2008]).

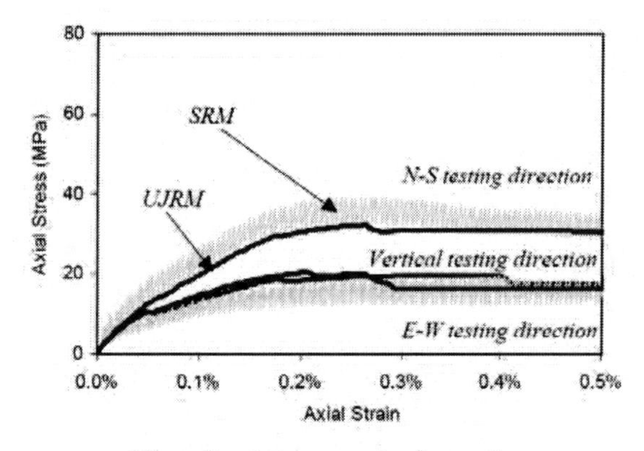

(a) Unconfined Compressive Strength

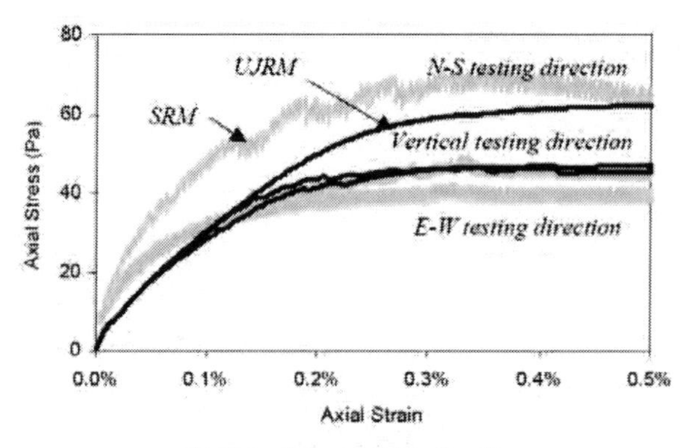

(b) Triaxial at 5 MPa Confinement

Figure 9.—Stress-strain plots from *(a)* uniaxial and *(b)* triaxial compression testing of carbonatite SRM samples from Palabora Mine. The SRM is compared to the continuum-based FLAC3D representation of the SRM results via the ubiquitous joint rock mass model (UJRM). The results show the significant anisotropy in the strength response [Sainsbury et al. 2008].

response in moduli and strength (Figures 8 and 9). The anisotropy in response is directly related to the vertical orientation of the fracture system.

The strength and modulus anisotropy determined from the SRM were taken into account in the FLAC3D continuum model by using a strain-softening, ubiquitous joint model in which the stochastic variability of the jointing, described by the DFN, is randomly seeded into the FLAC3D zones. The properties assigned to the intact rock and jointing are derived through the SRM strength testing such that the FLAC3D model reproduces the basic uniaxial and triaxial response (Figure 9). Whereas the typical application of the caving algorithm assumes isotropic rock mass properties, this particular approach provides for an orientation of weakness resulting from the orientation of the dominant jointing.

Sensitivity studies of the impact of the orientation of the anisotropy on generic cave growth and the mobilized rock mass for a given undercut area and total draw (3.9 Mt) within the FLAC3D caving algorithm were conducted [Sainsbury et al. 2008]. As shown in Figure 10, for the same amount of total draw, the associated zone of deformation (the mobilized zone based on >1 m of movement) is significantly different and affected by the assumed orientation of anisotropy. A horizontal weakness orientation promotes rapid cave growth vertically, while a vertical anisotropy results in minimal cave growth—in other words, the cave tends to grow at right angles to the direction of primary anisotropy. These examples are hypothetical since the anisotropy directions are uniform for each element (except for the isotropic case) and thus show an exaggerated impact of jointing. However, the point here is that, if strong anisotropy in the rock mass structure exists, it may significantly impact the direction and rate of growth of the cave.

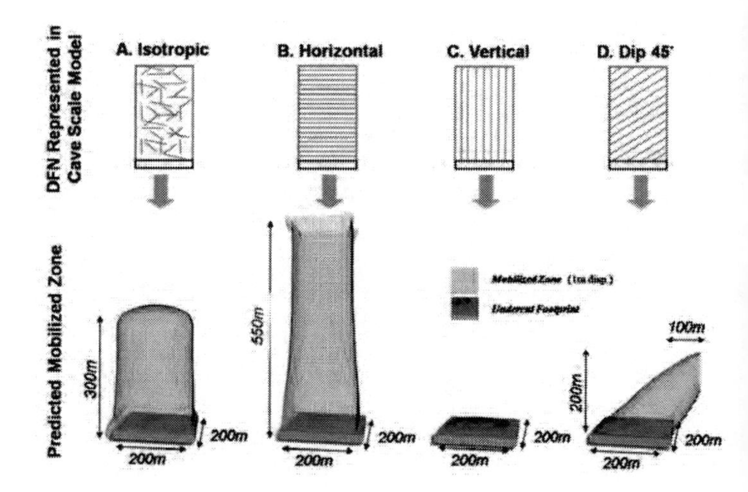

Figure 10.—Sensitivity of cave growth to the primary orientation of the fracturing [Sainsbury et al. 2008].

Application of the Caving Algorithm and SRM to Back Analysis of the Palabora Cave

A back analysis of the caving behavior of Rio Tinto's Palabora Mine was conducted as part of the MMT project as a validation exercise of the SRM methodology, and, in particular, the methodology for representing material anisotropy defined by the SRM testing, and was included in the FLAC3D model using the ubiquitous joint constitutive model. The following discussion summarizes the work of Sainsbury et al. [2008]. The back analysis was carried out by developing a large-scale FLAC3D model of the pit and underground and by advancing the undercut and pull at the draw rate specified by the actual draw schedule. The outputs of the model are the isosurfaces of the mobilized zone, the yield zone, and the estimated seismogenic zone. The seismogenic zone, which will precede the yield zone, is estimated by applying stresses in the FLAC3D elements to the following relation [Diederichs 1999], which is based on the threshold failure envelope when seismicity is initiated:

$$\sigma 1 = \sigma 3 + [0.2 \text{ to } 0.4] * UCSlab \quad (5)$$

The results of the analyses are shown in Figures 11 and 12. Figure 11 shows the isosurfaces of the mobilized and yield zones as a function of undercut advance from 2002 to 2004 (i.e., before and after initiation of the north wall slope failure). As seen in this figure, the predicted yield zone isosurface propagates rapidly up-dip and affects the stability of the pit bottom in advance of the caved (mobilized) zone. The stress concentration introduced at the pit bottom, coupled with the vertical stress relief due to removal of the pit material, causes the early interaction of the cave and pit. The failure of the pit toe at the north wall, induced by interaction with the cave, results in failure up-dip in the slope.

Verification of the model's ability to simulate the vertical advance and shape of the cave was obtained by

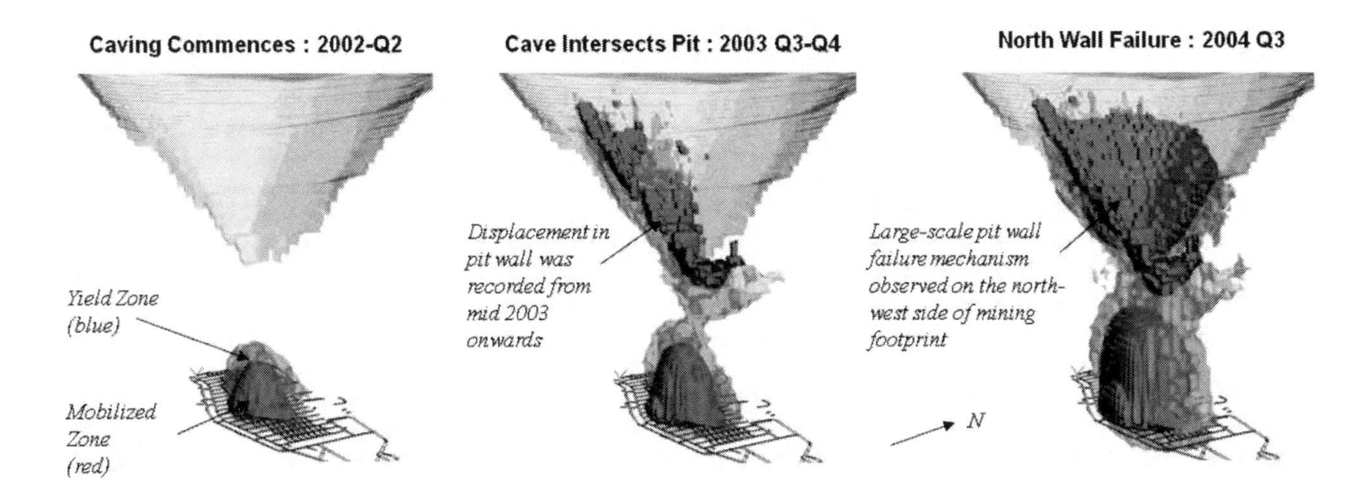

Figure 11.—Isometric view of FLAC3D model output showing progression of mobilized zone *(red)* and yield zone *(blue)* as undercut is advanced from 2002 to 2004 [Sainsbury et al. 2008].

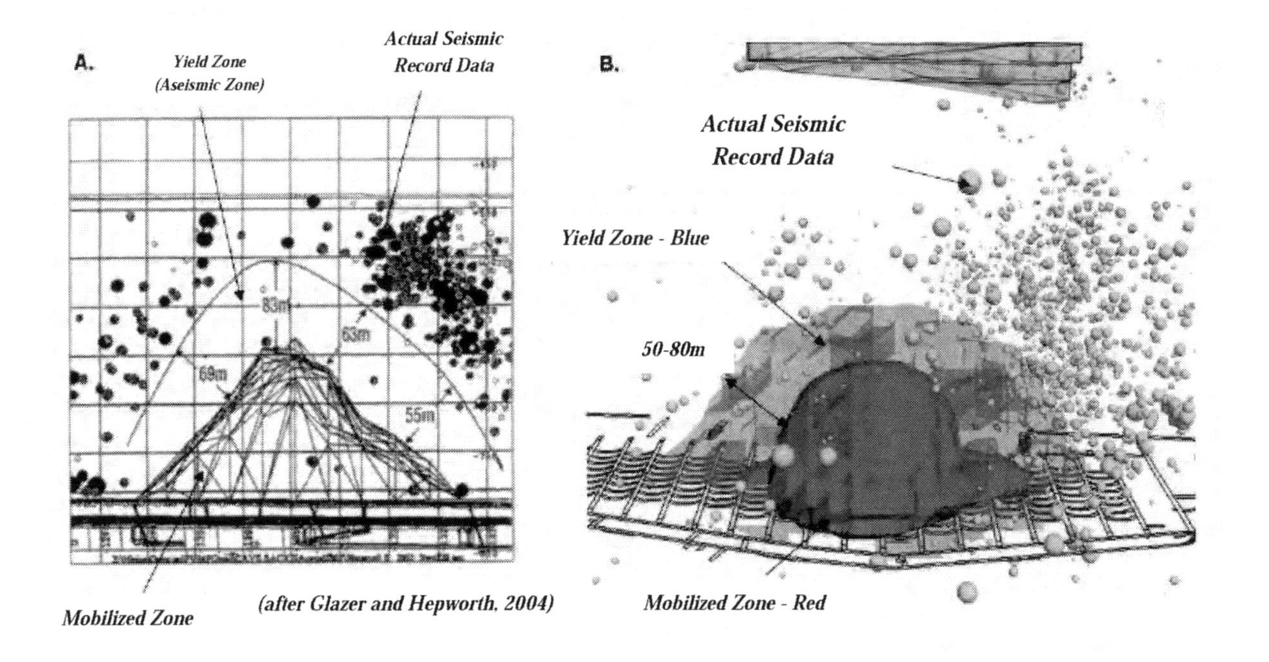

Figure 12.—Comparison of the location and dimensions of the aseismic and seismogenic zones from acoustic monitoring (*left,* Glazer and Hepworth [2004]) with FLAC3D predictions of the yield zone (*right,* Sainsbury et al. [2008]).

comparing the isosurfaces of the yield and seismogenic zones to measurements of acoustic emission made by the mine-wide microseismic system [Glazer and Hepworth 2004]. As shown in Figure 12B, the yield zone in the FLAC3D model is predicted to be approximately 50–80 m in extent beyond the mobilized zone. The yield zone is aseismic in nature since the higher-energy acoustic emissions are associated with the high-stress concentrations that have been shed in advance of the previously failed rock mass. The data gathered from the microseismic system indicate a similar region of yield in advance of the cave. The FLAC3D model also shows comparable (to the microseismic data) timing to the failure of the crown pillar between the pit floor and the shedding of stress concentration and seismicity to the confined region around the pit base and sidewalls.

CONCLUSION

This paper described a methodology for predicting cave growth that was developed over a number of years through a mining industry-funded research program—the MMT research project. The prediction of caving requires knowledge of the strain-softening behavior of the rock mass, since the prediction requires representation of the process of failure from the in situ rock mass to ultimate fragmentation of the ore body. Two methodologies were described to estimate strain-dependent strength properties of the rock mass for inclusion in a strain-softening, Mohr-Coulomb constitutive model. The first is the use of empirical relationships for rock mass properties using the GSI methodology of Hoek and parameters derived from back analysis of numerous caves. The second, more promising approach is the use of the SRM technique, in which a rigorous methodology is used to combine the material response of the basic geologic constituents (intact rock and fractures) to define the rock mass constitutive behavior. The caving algorithm described has been applied to back analysis (and forward analysis) at a significant number of caves and shows reasonable correlation to the basic parameters of interest in mining: the shape of the cave, the rate of cave growth as a function of draw, and the resultant cracking and subsidence regions.

REFERENCES

Brummer RK, Li H, Moss A [2006]. The transition from open pit to underground mining: an unusual slope failure at Palabora. In: Proceedings of the South African Institute of Mining and Metallurgy, International Symposium on Rock Slopes in Open Pit Mining and Civil Engineering, pp. 411–420.

Diederichs MS [1999]. Instability of hard rockmasses: the role of tensile damage and relaxation [Dissertation]. Waterloo, Ontario, Canada: University of Waterloo.

Duplancic P, Brady BH [1999]. Characterization of caving mechanisms by analysis of seismicity and rock stress. In: Proceedings of the Ninth International Congress on Rock Mechanics (Paris, France) (2):1049–1053.

Glazer S, Hepworth N [2004]. Seismic monitoring of block cave crown pillar: Palabora Mining Company, RSA. In: Proceedings of MassMin 2004 (Santiago Chile, August 22–25, 2004), pp. 565–569.

Hoek E, Brown ET [1997]. Practical estimates of rock mass strength. Int J Rock Mech Min Sci *34*(8):1165–1186.

Laubscher DH [2000]. Block caving manual. Prepared for the International Caving Study. JKMRC and Itasca Consulting Group, Brisbane, Australia.

Lorig L [2000]. Methodology and guidelines for numerical modelling of undercut and extraction-level behaviour in caving mines. Report to International Caving Study (1997–2000).

Mas Ivars D, Potyondy DO, Pierce M, Cundall PA [2008]. The smooth-joint contact model. In: Proceedings of the Eighth World Congress on Computational Mechanics and Fifth European Congress on Computational Methods in Applied Sciences and Engineering (Venice, Italy, June 30-July 5, 2008).

Pierce M, Young RP, Reyes-Montes J, Pettitt W [2006]. Six monthly technical reports, caving mechanics, subproject 4.2: research and methodology improvement; and subproject 4.3: case study application. Itasca Consulting Group report to Mass Mining Technology Project, 2004–2007, September.

Pierce M, Cundall P, Potyondy D, Mas Ivars D [2007]. A synthetic rock mass model for jointed rock mechanics. In: Eberhardt E, Stead D, Morrison T, eds. Rock mechanics: Meeting society's challenges and demands. London: Taylor and Francis, pp. 341–349.

Sainsbury B, Pierce ME, Mas Ivars D [2008]. Analysis of caving behaviour using a synthetic rock mass: ubiquitous joint rock mass modelling technique. In: Potvin Y, Carter J, Dyskin A, Jeffrey R, eds. Proceedings of the First Southern Hemisphere International Rock Mechanics Symposium (SHIRMS 2008) (September 16–19, 2008). Perth, Australia: Australian Centre for Geomechanics, Vol. 1, pp. 243–254.

Serafim JL, Pereira JP [1983] Considerations on the geomechanical classification of Bieniawski. In: Proceedings of the International Symposium on Engineering Geology and Underground Construction (Lisbon, Portugal), (1):II.33-II.42. Lisbon, Portugal: Sociedade Portuguesa de Geotécnica/Laboratório Nacional de Engenharia Civil.

CHARACTERIZATION OF NATURAL FRAGMENTATION USING A DISCRETE FRACTURE NETWORK APPROACH AND IMPLICATIONS FOR CURRENT ROCK MASS CLASSIFICATION SYSTEMS

By D. Elmo,[1] S. Rogers,[1] and D. Kennard[1]

ABSTRACT

Design practice in rock engineering relies significantly on a consistent characterization of the natural rock fracture network. For instance, an accurate representation of natural fragmentation is key to establishing representative rock mass material strength. The use of a Discrete Fracture Network (DFN) approach integrated with a Rock Block Analysis is shown to provide a powerful tool for studying the distribution of rock blocks as a function of discontinuity persistence, spacing, and orientation. This paper presents the results of a study to evaluate the fundamental factors influencing natural fragmentation, emphasizing the role of discontinuity geometry, orientation, and intensity. By providing a quantitative assessment of the in situ block size, the proposed approach is shown to have major implications for the characterization of the strength of rock masses with nonpersistent joints. These preliminary results form the basis for the development of an innovative DFN-based rock mass characterization approach, which combines empirically derived rock mass properties with orientation-dependent parameters to account for rock mass anisotropy.

INTRODUCTION

Numerical modeling has played and will continue to play an important role in rock engineering design [Hoek et al. 1990]. The scope of a model is not to represent rock engineering processes in their entirety. Rather, the objective of the analyst is to determine which process needs to be considered explicitly and which can be represented in an average way. Parametric characterization and its association with sample size, representative elemental volume, and homogenization/upscaling represent fundamental problems faced in realistic modeling. For this reason, any modeling and subsequent rock engineering design will, by necessity, include some component of subjective judgment.

Recent advances in the field of data capture and synthesis have allowed the derivation of more accurate three-dimensional (3-D) models of naturally jointed rock masses, overcoming some of the limitations inherent in an infinite ubiquitous joint approach. The true discontinuous and inhomogeneous nature of the rock mass should be reflected in most modeling conceptualization. Thus, the importance of discontinuity persistence cannot be overemphasized if realistic characterization and fracture analysis is to be undertaken [Kalenchuk et al. 2006; Kim et al. 2007; Elmo et al. 2007, 2008]. It is clear that the volume, shape, and stability of rock blocks depend on the characteristics of the natural rock fracture network. In this context, the Discrete Fracture Network (DFN) approach represents an ideal numerical tool with which to synthesize realistic fracture network models from digitally and conventionally mapped data.

Whereas numerical simulations provide a potentially useful means of overcoming some of the limits of the empirical methods, empirical approaches such as rock mass characterization and classification systems still represent a fundamental component for many applications in both mining and rock engineering practice. Ideally, both qualitative and quantitative data should be collected as part of the rock mass characterization process, providing the necessary parameters for a subsequent classification analysis. Fundamental aspects of rock mass characterization include: definition of an accurate geological model, geotechnical data collection, assessment of the role of major geological structures, and determination of rock mass properties.

This paper introduces the preliminary concepts of a DFN-based rock mass characterization approach, which couples empirically derived mechanical properties with orientation-dependent parameters to account for rock mass anisotropy. This set of properties can then be incorporated into a continuum finite-element or finite-difference model. The objective is to provide an improved link between mapped fracture systems and rock mass strength compared to the current practice of using empirical rock mass classifications alone.

ROCK MASS CHARACTERIZATION METHODS AND ROCK MASS PROPERTIES

One of the key aspects of rock engineering is establishing representative rock mass material strength and deformability characteristics. This section briefly summarizes available approaches for deriving the mechanical properties of a jointed rock mass.

[1]Golder Associates Ltd., Greater Vancouver Office, Burnaby, British Columbia, Canada.

Rock Mass Characterization Methods

Traditionally, rock mass characterization has been achieved using empirical classification methods, including the Geological Strength Index (GSI) [Hoek et al. 1995, 2002], the Rock Mass Rating (RMR) system [Bieniawski 1989], and the Q-index [Barton et al. 1974]. These classification systems are useful tools that (1) identify significant parameters influencing rock mass behavior, (2) derive quantitative data for engineering design, and (3) provide a quantitative measure to compare geological conditions at different sites. In different ways within these classification systems, various discontinuity properties such as discontinuity spacing, Rock Quality Designation (RQD), and discontinuity roughness are weighted and combined to give a value (or range of values) describing the rock mass characteristics. Rock mass classification systems such as the RMR, the Q-index, or the coupled GSI/Hoek-Brown approach are traditionally used to derive properties for numerical analysis of rock engineering problems. A summary of RMR, Q, and GSI-based empirical relationships for estimating mechanical properties of the rock mass is presented in Table 1.

Table 1.—Empirical relationships for deriving estimates of basic rock mass strength and deformability properties based on the RMR, GSI, and Q rock mass classification systems (modified from Vyazmensky [2008])

RMC System	Estimates of Rock Mass Strength and Deformability Characteristics	Reference
RMR	$E_m = 10^{(RMR-10)/40}$ (GPa) $\phi = 5 + RMR/2$ $c = 5RMR$ (kPa)	Serafim & Pereira (1983) Bieniawski (1989)
Q	$E_m \approx 10 \times Q_c^{1/3}$ (GPa) $"\phi" \approx \tan^{-1}\left(\frac{J_r}{J_a} \times \frac{J_w}{1}\right)$ $"c" \approx \left(\frac{RQD}{J_n} \times \frac{1}{SRF} \times \frac{\sigma_c}{100}\right)$ (MPa) where: $Q_c = Q(\sigma_c/100)$ - normalized Q; σ_c – uniaxial compressive strength (MPa)	Barton (2002)
GSI	$E_m = E_i\left(0.02 + \frac{1-D/2}{1+e^{((60+15D-GSI)/11)}}\right)$ (MPa) $\phi' = a\sin\left[\frac{6am_b\left(s+m_b\sigma'_{3n}\right)^{a-1}}{2(1+a)(2+a)+6am_b\left(s+m_b\sigma'_{3n}\right)^{a-1}}\right]$ $c' = \frac{\sigma_{ci}\left[(1+2a)s+(1-a)m_b\sigma'_{3n}\right]\left(s+m_b\sigma'_{3n}\right)^{a-1}}{(1+a)(2+a)\sqrt{1+\left(6am_b\left(s+m_b\sigma'_{3n}\right)^{a-1}\right)/((1+a)(2+a))}}$ (MPa) where: E_i – intact rock Young's modulus; D - disturbance factor; a, s, m_b – material constant; $\sigma_{3n} = \sigma'_{3max}/\sigma_{ci}$, σ'_{3max} - upper limit of confining stress	Hoek et al. (2002) Hoek & Diederichs (2006)

Numerical Methods

Numerical modeling methods are increasingly being used for deriving rock mass properties through simulated rock mass tests. For instance, the Synthetic Rock Mass Approach proposed by Pierce et al. [2007] and Cundall et al. [2008] involves constructing a synthetic sample of the rock mass in two or three dimensions by bonding together thousands of circular or spherical particles. DFN-based preinserted joints are introduced by debonding particles along specific joint surfaces and employing a sliding joint model. Similarly, by coupling a DFN model with a hybrid finite-discrete analysis, Elmo and Stead [2009] have demonstrated that the use of a synthetic approach could allow engineers to simulate the rock mass response to loading by fully accounting for existing jointing conditions while also explicitly accounting for size and shape (scale) effects.

Important factors that play a major role in the development of synthetic rock mass properties (independently of the adopted numerical methodology) include the following:

(1) The scale of the synthetic rock mass tests must be sufficiently large to capture the representative elemental volume for local jointing conditions and, in order to accommodate structural anisotropy, should be repeated to allow for different orientations of σ_1 and σ_3 relative to the joint orientations.

(2) The stochastic nature of the embedded DFN model is such that there are an infinite number of possible realizations of the 3-D fracture system based on the mapped data. Accordingly, the analysis should incorporate a Monte Carlo approach, by running a certain number of DFN models and incorporating those in the subsequent synthetic rock mass modeling. The necessary large computing times required to perform such an analysis could limit the effectiveness of the Synthetic Rock Mass Approach and the associated definition of specific formulations of rock mass behavior.

(3) Most geomechanical models rely on a form of calibration process. For lab-scale intact rock samples, the calibration process may be a straightforward procedure. Instead, calibrating a model against the strength of a large-scale and naturally fractured rock mass may not be directly possible unless considering large-scale physical testing. References to large-scale testing can be found in the literature, but these are usually limited. Some researchers (e.g., Hoek and Brown [1980]) have proposed some forms of upscaling relationships for estimating the strength of larger intact rock samples, but these are typically limited

to rock specimens with a diameter not greater than 500 mm.

THE DFN APPROACH AS A TOOL TO SIMULATE MORE REALISTIC GEOLOGICAL MODELS

Rock discontinuities can be characterized in terms of their orientation, intensity, and spatial distribution, in addition to their strength and deformability. With the exception of fully explicit modeling of an individual fracture or simplified fracture sets, the use of a stochastic DFN approach provides the best option for creating realistic geometric models of fracturing, reflecting the heterogeneous nature of a specific fractured rock mass. The basis of DFN modeling is the characterization of each discontinuity set within a structural domain using statistical distributions to describe variables such as orientation, persistence, and spatial location of the discontinuities. The DFN approach maximizes the use of discontinuity data from mapping of exposed surfaces and boreholes or any other source of spatial information. Discontinuity data sampled from exposures in variably oriented outcrops (two-dimensional) and boreholes (one-dimensional) can be used to synthesize a 3-D stochastic discontinuity model that shares the statistics of the samples and allows for the incorporation of specific (deterministic) discontinuities such as larger mappable structures. Observations of exposed rock faces, at or near the project site, have the advantage of allowing direct measurements of discontinuity orientation, spacing, and persistence and the identification of discontinuity sets. Other large-scale geometrical and structural features can be readily observed. Increasingly, digital photogrammetry and laser scanning techniques (LiDAR) provide an alternative method for surface fracture characterization.

The typical process involved in the generation of a DFN model includes the definition of (1) a fracture spatial model, (2) fracture orientation distribution, (3) fracture terminations, (4) fracture radius distribution, and (5) fracture intensity. Validation of the DFN model is achieved by comparing the orientation, intensity, and pattern of the simulated fracture traces with those measured in the field using a simulated sampling methodology.

The proprietary code FracMan [Dershowitz et al. 1998; Golder Associates 2009] is the platform used in the current study for data analysis and synthesis. FracMan allows the 3-D visualization of blocks defined by intersecting discontinuities in the DFN model by employing either an ***implicit*** cell mapping algorithm or a more conventional ***explicit*** block search algorithm [Dershowitz and Carvalho 1996]. Whereas the latter provides an accurate estimate of block shape and volume, its use is better suited for the kinematic assessment of block stability. The cell mapping algorithm is optimized to provide an initial estimate of the rock natural fragmentation. As shown in Figure 1, the cell mapping algorithm works by initially identifying all of the fracture intersections with the specified grid elements. This results in a collection of grid faces and connection information, which is then used to construct a rock block of contiguous grid cells.

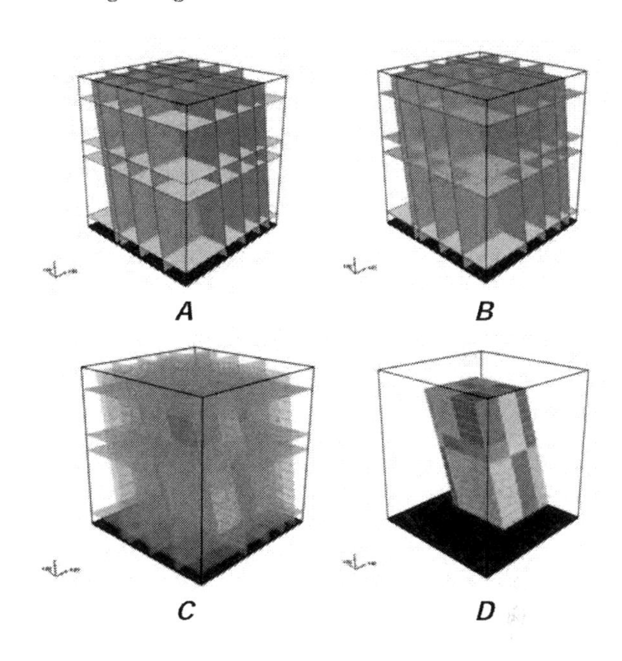

Figure 1.—Cell mapping algorithm: (*A*) initial DFN, (*B*) fractures are mapped to the specified grid, (*C*) regular blocks are formed along the grid cells, and (*D*) final rock block model.

MODELING NATURAL FRAGMENTATION USING AN INTEGRATED DFN-ROCK BLOCK ANALYSIS APPROACH

It is safe to assume that the number of blocks formed and both their volume and relative continuity within the geological model are strongly related to the geometry and number of intact rock bridges. If joints were assumed to be fully persistent, this would erroneously lead to more removable blocks than actually existed in situ. With the exception of size and geometry of the excavation, which are fixed, fracture orientation and length are expressed within the DFN model as probability density functions. The accuracy of block predictions clearly depends on the precision of the initial descriptive parameters, including the cell grid size used in the cell mapping algorithm.

The current study builds on the initial work by Elmo et al. [2008]; the principal difference consists in the integration of the cell mapping algorithm for fragmentation analysis. A series of conceptual discrete-fracture networks was generated in FracMan using the parameters listed in Table 2.

The results (Figure 2) clearly show that, for a simple fracture network with three joint sets, fracture length has a dramatic impact on the block forming potential for a given linear intensity P_{10} assuming the equivalence of all other properties. These results are in agreement with observations by Chan and Goodman [1987] and Hoerger and Young [1990]. The analysis shows that there is a decrease in the number of rock blocks, independently of fracture length, for models with relatively wider spaced fractures (i.e., decreasing linear intensity P_{10}). Relatively short fractures coupled with a relatively high-intensity P_{10} can still produce a block assemblage characterized by very few and widely spaced intact rock bridges. Clearly, this would have major implications with respect to rock mass strength, the development of primary fragmentation, and the mobilization of rock blocks. By interpreting the results with respect to block volume and number of blocks (calculated as the mean of all sequential stages) (Figure 2*A*), a preliminary attempt has been made to characterize the degree of natural fragmentation as a function of fracture length and fracture intensity P_{10} (Figure 2*B*).

Table 2.—Parameters used in the conceptual DFN-Rock Block Analysis

(NOTE: Each DFN model is generated within a 10-m by 10-m by 10-m box region.)

Parameter	Set 1	Set 2	Set 3
Orientation (°)	000/90	090/90	000/00
Fracture length (m)	1, 2, 5, and 10	1, 2, 5, and 10	1, 2, 5, and 10
Linear fracture intensity P_{10} $^{(m-1)}$	1, 2, 3, and 4	1, 2, 3, and 4	1, 2, 3, and 4

For a given fracture length and fracture intensity, the current DFN analysis demonstrates that the massive to blocky character of a hypothetical rock mass could quantitatively and qualitatively be expressed as a function of mean block volume and number of fully formed blocks. In Figure 3, the massive to blocky character is defined according to the quantitative method to assist in the use of the GSI system for rock mass classification proposed by Cai et al. [2004]. As discussed by Cundall et al. [2008], this method is not scale independent, and the block volumes suggested by Cai et al. [2004] should be

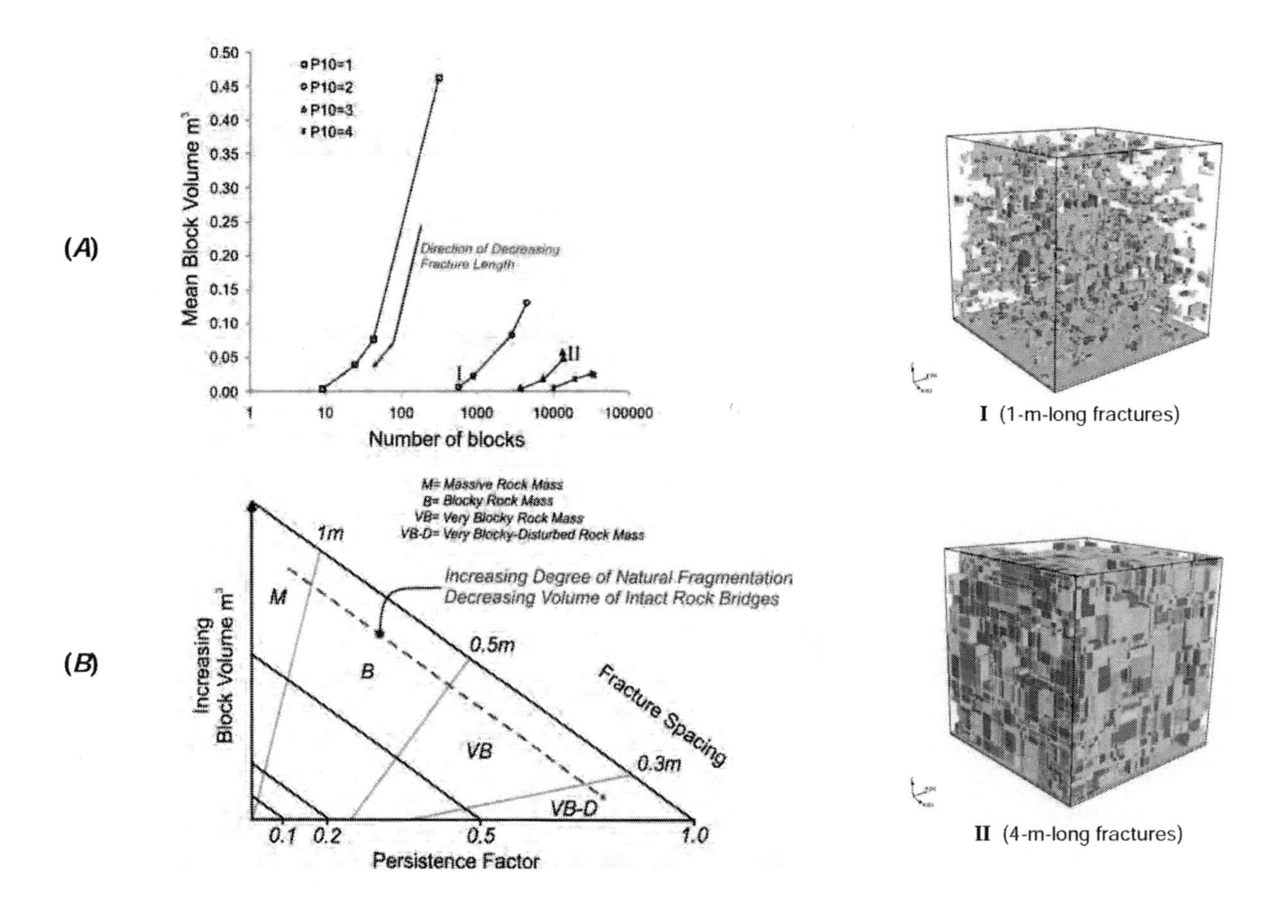

Figure 2.—(*A*) Variation of block volume and number of blocks as a function of fracture intensity P_{10} with visualization of blocky rock masses at points I and II, respectively, and (*B*) qualitative characterization of the degree of natural fragmentation as a function of persistence factor and fracture spacing (modified from Elmo et al. [2008]).

interpreted relative to the scale of a tunnel. To account for that, the current DFN-based fragmentation analysis is carried out within a 10-m by 10-m by 10-m region.

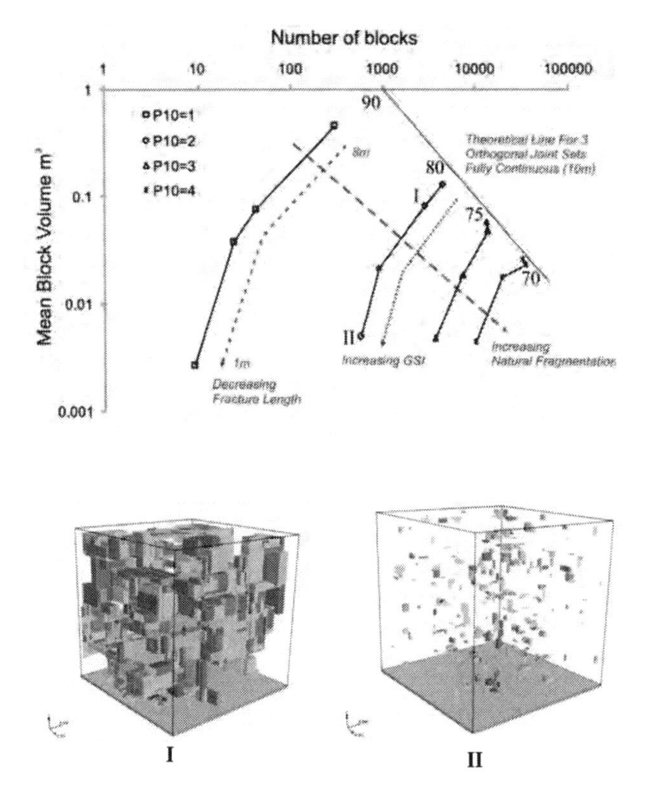

Figure 3.—Rock Block Analysis carried out for a 10-m by 10-m by 10-m region volume. Quantitative characterization of the rock mass natural fragmentation is expressed as a function of mean block volume, number of fully formed blocks, and GSI.

A DFN-BASED ROCK MASS CHARACTERIZATION APPROACH

The use of numerical modeling to derive synthetic rock mass properties is an elegant solution to the problem (establishing representative rock mass properties) long recognized as one of the main challenges in rock mechanics. However, at present it is not practical to perform 3-D large-scale engineering analysis within discontinuum codes that are capable of simulating intact rock fracturing. Typically, numerical tests of rock mass behavior are used in conjunction with continuum codes.

Work is currently ongoing to provide an alternative modeling route for estimating rock mass properties for continuum modeling. Because any process of deriving numerically simulated rock mass properties ultimately requires the explicit use of a DFN embedded within an intact rock matrix, it is proposed to use the DFN model itself, coupled with empirically derived rock mass properties, to provide the necessary input parameters for the continuum analysis. By superimposing a grid structure to the DFN model, the proposed approach takes advantage of existing rock mass classification schemes to obtain, for each grid cell, an equivalent rock mass rating (Figure 4). Using the relationships in Table 1 is then possible to derive an initial estimate of rock mass properties, including rock mass cohesion and friction. Modeling of anisotropic behavior would be indirectly undertaken within a continuum code by assigning ubiquitous joint orientations for specific zones in the model. Work is ongoing to define a sampling procedure for each grid cell in the DFN model to define the ubiquitous joint dip, dip direction, and size.

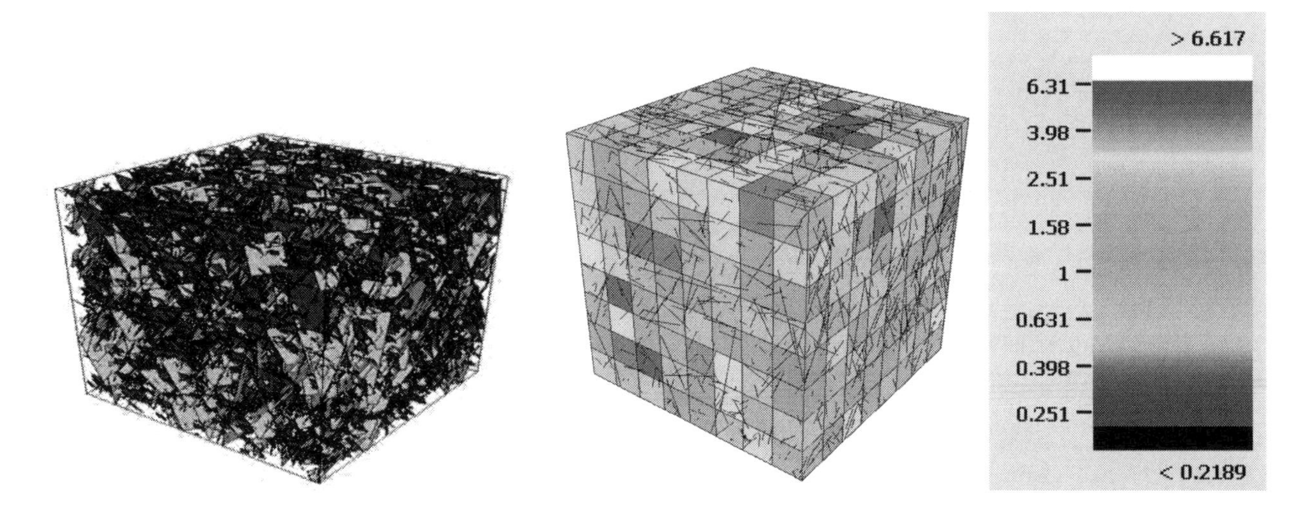

Figure 4.—DFN model and grid cell visualization of a sampled parameter (volumetric intensity P_{32} is shown). Logarithmic scale, darker shades define zones with relatively lower volumetric fracture intensity.

Table 3.—Parameters used for the DFN: fragmentation analysis and correlation with RQD data

Parameter	Set 1	Set 2	Set 3	Set 4
Subset A:				
Orientation (°)	203/82	119/49	327/74	277/86
Mean fracture length[1] (m)	5	5	5	5
Linear fracture intensity P_{10}	1, 2, 4, and 8	1, 2, 4, and 8	1, 2, 4, and 8	1, 2, 4, and 8
Subset B:				
Orientation (°)	203/82	119/49	327/74	277/86
Mean fracture length[1] (m)	30	30	30	30
Linear fracture intensity P_{10}	0.25	0.25	0.25	0.25

[1]Negative exponential distribution was used.

A PRELIMINARY STUDY ON THE USE OF A DFN METHOD TO PROVIDE ESTIMATES OF ROCK MASS PROPERTIES

It is possible to correlate the results of a combined DFN-Rock Block Analysis with existing rock mass classification systems. The scope of the current DFN modeling is to further develop these correlations, investigating whether inputs to classification schemes can be expressed as functions of specific DFN parameters.

The preliminary analysis considered a DFN model generated using the parameters listed in Table 3. The number of joint sets is defined using the J_n parameter convention (Q-index), and each set is defined using a function that randomly generates numbers within the range [0, 360] for dip direction and [0, 90] for dip. The dip and dip direction of the random joints are also defined using a similar function (Figure 5*A*). Each DFN model is generated within a 80-m by 80-m by 80-m box region, and three simulated boreholes are used to estimate the RQD of the rock mass (Figure 5*B*). The subsequent fragmentation analysis is carried out within a 10-m by 10-m by 10-m region (assumed pillar region).

The RQD provides a measure of rock mass quality from drill core and is used as an input into geomechanical classification schemes, mainly the RMR and the Q system. The results (Figure 6*A*) show that there is an apparent correlation between RQD and the DFN volumetric intensity P_{32}. The correlation is independent of orientation (i.e., number of joint sets). Notwithstanding the arbitrary choice of the input parameters of the current DFN model, there is an agreement between simulated results and results obtained for a DFN developed from an actual mine site. In Figure 6*B*, the total volume of formed blocks is divided by the total volume of the assumed pillar region to define a normalized volume index. The results show an apparent relationship between RQD and the degree of natural fragmentation of the assumed rock mass. It is argued that RQD is direction-dependent, and its value may change significantly with the azimuth and plunge of the simulated boreholes. There is a need to extend the analysis to consider different combinations of the assumed initial parameters used to generate the DFN models; this analysis is currently ongoing.

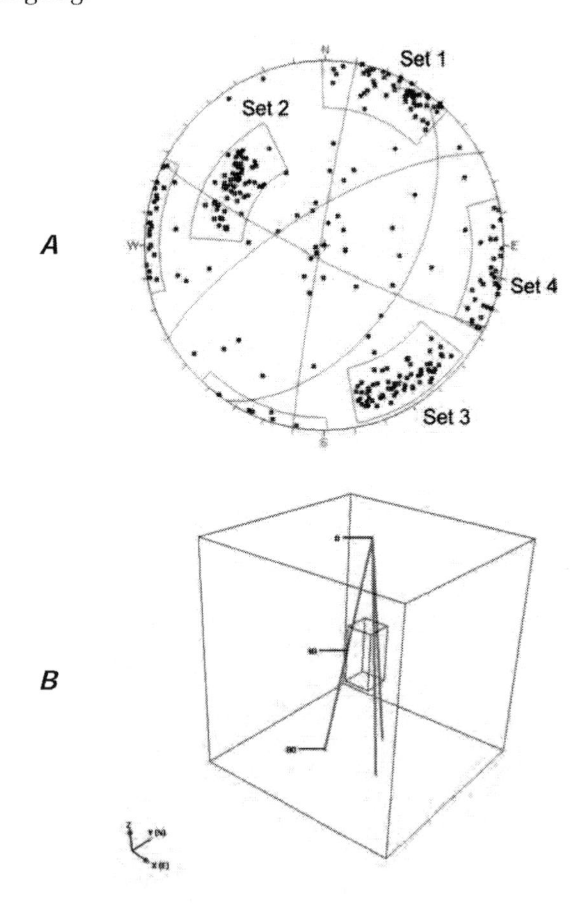

Figure 5.—(*A*) Stereoplot showing the four main joint sets and random joints; (*B*) boreholes used to calculate the RQD value for the simulated rock mass.

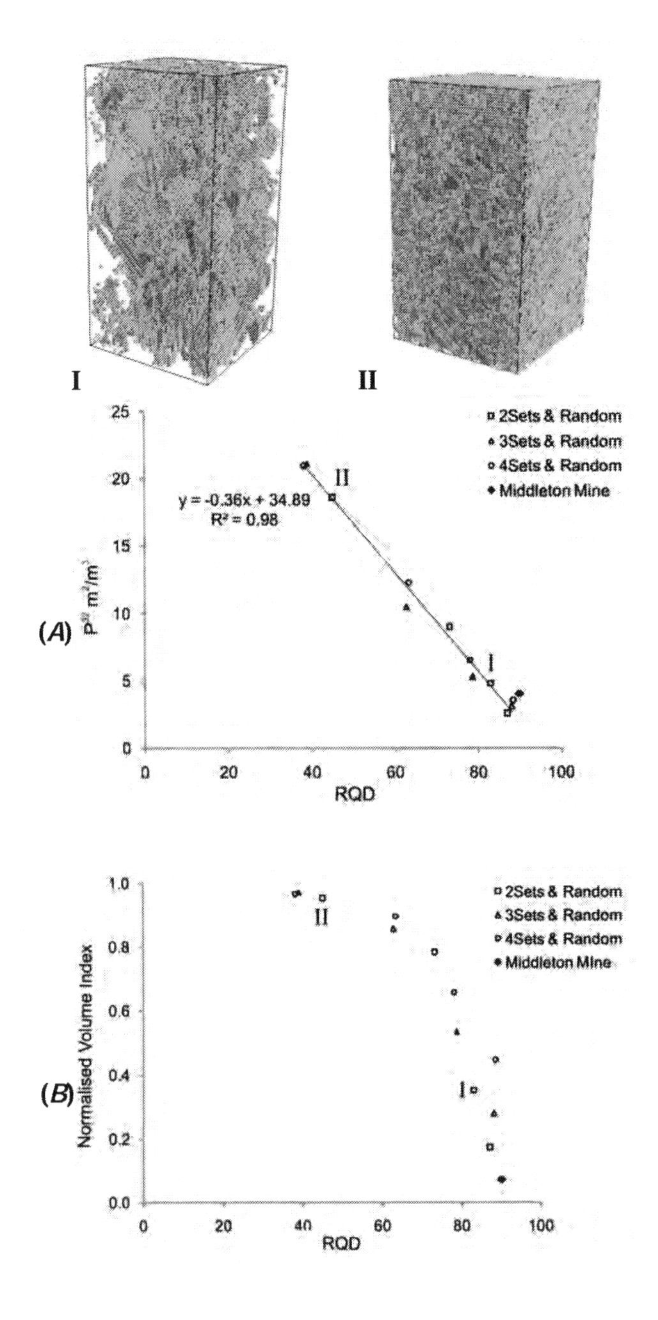

Figure 6.—Simulated relationships between RQD and (*A*) volumetric intensity P_{32} and (*B*) normalized volume index.

DISCUSSION AND CONCLUSIONS

A technique is being developed that would allow engineers to employ a DFN-based rock mass characterization approach to derive rock mass properties to be used in conjunction with continuum codes. The main advantage of the approach is that it relies on quantifiable field rock mass descriptors (fracture orientation, length, and intensity) and the results of a large number of DFN models can be quickly compared to provide an initial estimate of rock mass strength, accounting also for scale and anisotropy effects. Whereas the initial modeling shows encouraging results, research is currently ongoing to further extend and validate the proposed methodology by considering different combinations of fracture orientation, length, spacing, and termination modes.

REFERENCES

Barton N [2002]. Some new Q value correlations to assist in site characterization and tunnel design. Int J Rock Mech Min Sci *39*(2):185–216.

Barton N, Lien R, Lunde J [1974]. Engineering classification of rock masses for the design of rock support. Rock Mech *6*:189–236.

Bieniawski ZT [1989]. Engineering rock mass classification. New York: John Wiley & Sons.

Cai M, Kaiser PK, Uno H, Tasaka Y, Minami M [2004]. Estimation of rock mass deformation modulus and strength of jointed hard rock masses using the GSI system. Int J Rock Mech Min Sci *41*:3–19.

Chan LY, Goodman RE [1987]. Predicting the number and dimensions of key blocks of an excavation using block theory and joint statistics. In: Farmer IW, Daemen JJK, Desai CS, Glass CE, Neuman SP, eds. Proceedings of the 28th U.S. Symposium on Rock Mechanics (Tucson, AZ), pp. 81–87.

Cundall PA, Pierce ME, Mas Ivars D [2008]. Quantifying the size effect of rock mass strength. In: Proceedings of the First Southern Hemisphere International Rock Mechanics Symposium (Perth, Australia, September 16–19, 2008).

Dershowitz W, Carvalho J [1996]. Key-block tunnel stability analysis using realistic fracture patterns. In: Aubertin M, ed. Rock Mechanics Tools and Techniques. Proceedings of the Second North American Rock Mechanics Symposium: NARMS '96 (Montreal, Quebec, Canada, June 19–21, 1996). Rotterdam, Netherlands: A. A. Balkema, pp. 1747–1751.

Dershowitz W, Lee G, Geier J, LaPointe PR [1998]. FracMan: interactive discrete feature data analysis, geometric modelling and exploration simulation; user documentation. Seattle, WA: Golder Associates, Inc.

Elmo D, Stead D [2009]. An integrated numerical modelling-discrete fracture network approach applied to the characterisation of rock mass strength of naturally fractured pillars. Rock Mech Rock Eng (online).

Elmo D, Yan M, Stead D, Rogers S [2007]. The importance of intact rock bridges in the stability of high rock slopes: towards a quantitative investigation using an integrated numerical modelling-discrete fracture network approach. In: Proceedings of the International Symposium on Rock Slope Stability in Open Pit Mining and Civil Engineering (Perth, Australia, September 12–14, 2007).

Elmo D, Stead D, Rogers S [2008]. A quantitative analysis of a fractured rock mass using a discrete fracture network approach: characterization of natural fragmentation and implications for current rock mass classification systems. In: Proceedings of the Fifth International Conference and Exhibition on Mass Mining (Luleå, Sweden, June 9–11, 2008).

Golder Associates [2009]. FracMan 7.10. Redmond, WA: Golder Associates, Inc., FracMan Technology Group [http://fracman.golder.com].

Hoek E Brown ET [1980]. Underground excavations in rock. London: Inst Min Metall.

Hoek E, Diederichs MS [2006]. Empirical estimation of rock mass modulus. Int J Rock Mech Min Sci *43*:203–215.

Hoek ET, Grabinsky MW, Diederichs MS [1990]. Numerical modelling for underground excavation design. Trans Inst Min Metall. Section A: Mining Industry *100*:A22–A30.

Hoek ET, Kaiser PK, Bawden WF [1995]. Support of underground excavations in hard rock. Rotterdam, Netherlands: A. A. Balkema.

Hoek ET, Carranza-Torres CT, Corkum B [2002]. Hoek-Brown failure criterion – 2002 edition. In: Proceedings of the North American Rock Mechanics Society Conference (Toronto, Ontario, Canada).

Hoerger SF, Young DS [1990]. Probabilistic prediction of keyblock occurrences. In: Hustrulid WA, Johnson GA, eds. Rock mechanics contributions and challenges. Rotterdam, Netherlands: Balkema, pp. 229–236.

Kalenchuk KS, Diederichs MS, McKinnon S [2006]. Characterizing block geometry in jointed rock masses. Int J Rock Mech Min Sci *43*:1212–1225.

Kim BH, Cai M, Kaiser PK, Yang HS [2007]. Estimation of block sizes for rock masses with non-persistent joints. Rock Mech Rock Eng *40*:145–168.

Pierce M, Cundall P, Potyondy D, Mas Ivars D [2007]. A synthetic rock mass model for jointed rock. In: Proceedings of the First Canada-U.S. Rock Mechanics Symposium (Vancouver, British Columbia, Canada, May 27–31, 2007), Vol. I, pp. 341–349.

Serafim JL, Pereira JP [1983]. Consideration of the geomechanic classification of Bieniawski. In: Proceedings of the International Symposium on Engineering Geology and Underground Construction, pp. 1133–1144.

Vyazmensky A [2008]. Numerical modelling of surface subsidence associated with block caving using a finite-element/discrete-element approach [Dissertation]. Vancouver, British Columbia, Canada: Simon Fraser University.

THREE-DIMENSIONAL MODELING OF LARGE ARRAYS OF PILLARS FOR COAL MINE DESIGN

By Gabriel S. Esterhuizen, Ph.D.,[1] and Christopher Mark, Ph.D.[1]

ABSTRACT

The stability of the pillar line during retreat pillar mining is affected by the mining sequence, the mining geometry and the properties of the pillars, the gob, and the surrounding strata. Numerical models can assist in quantifying the complex interaction between these components as pillars are extracted and the roof caves to form the gob. However, modeling the details of the pillar geometry, as well as the large-scale surrounding strata, in a single three-dimensional model can pose significant challenges in terms of computer memory and solution run times.

This paper describes a modeling technique that allows large arrays of pillars to be modeled by making use of equivalent elements that capture the stress-strain response of the pillars and the immediate roof and floor rocks. The stress-strain response is obtained from numerical models that have been calibrated against instrumented case studies. The pillar response is programmed into relatively large equivalent elements in a large-scale three-dimensional model, negating the need to model the details of the pillars and surrounding excavations.

An example is presented in which this method is used to assess retreat mining in two different geological settings. This modeling technique significantly improves the capabilities for evaluating retreat mining pillar stability in a variety of geotechnical conditions.

INTRODUCTION

Background

On August 6, 2007, a violent coal bump occurred at the Crandall Canyon Mine near Price, UT. Six miners working at the time of the incident were presumed trapped. Ten days later, three rescuers were killed in a second bump. Rescue efforts were suspended, and the original six miners remained entrapped and were presumed to have been fatally injured.

The miners at Crandall Canyon had been engaged in the process of pillar recovery when the disaster occurred. In the United States, pillar recovery accounts for no more than 10% of the coal mined underground, yet historically it has been associated with more than 25% of all ground fall fatalities [Mark et al. 2003]. Maintaining "global stability" through proper pillar design is essential to safe pillar recovery [Mark and Zelanko 2005]. The Mine Safety and Health Administration (MSHA) investigation report concluded that the Crandall Canyon disaster occurred because the pillars were too small to carry the overburden loads [Gates et al. 2008].

Current Methods of Design

Retreat pillar mining presents a complex problem for engineering analysis and design. Any analysis is required to account for the three-dimensional characteristics of the overall panel layout, pillar loading and yield, the stability of the rooms, caving of the roof rocks after pillars have been extracted, the impact of variable strength strata, and variable field stresses. In addition, the surrounding strata are layered sedimentary rocks with highly anisotropic strength and deformation characteristics.

Adding to the complexity, pillar recovery is conducted in a variety of geologic environments using a range of mining methods. In the Western United States, the terrain is extremely rugged, the overburden consists largely of thick, strong sandstones, and mining is typically conducted at depths that can exceed 600 m. In the northern Appalachian coalfields, the topography is rolling, the rocks are weaker, and the typical cover depth is less than 300 m. Conditions typically fall between these two extremes in the central and southern Appalachian coalfields, where most pillar recovery operations are located.

The complexity of the problem led the National Institute for Occupational Safety and Health (NIOSH) to develop an empirical method for pillar design called the Analysis of Retreat Mining Pillar Stability (ARMPS) [Mark and Chase 1997]. The main strength of ARMPS is that it relies upon a large database of actual mining case histories to suggest the proper stability factors under different circumstances. The original database of 150 retreat mining case histories was later updated with nearly 100 more from mines operating at depths in excess of 225 m [Chase et al. 2002].

ARMPS has been used extensively to design pillars and to evaluate roof control plans in the central Appalachian coalfields for nearly a decade. More recently, NIOSH developed the Analysis of Multiple-Seam Stability (AMSS) program, which extends ARMPS to multiple-seam situations [Mark et al. 2007]. In the wake of the Crandall Canyon disaster, MSHA issued a Program Information Bulletin [Stricklin and Skiles 2008] and a Procedure Instruction Letter [Skiles and Stricklin 2008] that

[1]Pittsburgh Research Laboratory, National Institute for Occupational Safety and Health, Pittsburgh, PA.

essentially require that ARMPS be used in all roof control plan evaluations to help ensure that pillars are properly designed.

ARMPS uses relatively simple models to estimate the strength of the pillars and the magnitudes of the loads applied to them [Mark and Chase 1997]. For the pillar strength, ARMPS uses the Mark-Bieniawski formula [Mark 1987]. Tributary area equations are used to estimate the pillar development loads, while the "abutment angle" concept is used for the loads transferred to pillars during pillar extraction. Analysis of the deep-cover case histories in the ARMPS database indicated that the loading model may be less accurate for mining geometries that are highly "subcritical" (i.e., where the depth is much greater than the panel width) or when bridging of strong strata may occur in the overburden. In such cases, it seems that ARMPS may overestimate the loads applied to the panel pillars while underestimating the load carried by the interpanel barrier pillars.

Numerical models have found limited application for evaluating retreat pillar mining owing to the excessive demands of computing hardware and model run times that are required to realistically simulate the complexities of the problem. One approach has been to simplify the problem through the use of boundary-element methods, in which only the coal seam(s) are modeled and the surrounding strata are assumed to be homogeneous. An example of such a program is LaModel [Heasley 1997], which uses the thin-plate formulation of the boundary-element method [Salamon 1991]. The program was originally developed by NIOSH and has found wide application in the U.S. coal mining industry. Pillar yield and gob compaction are modeled by implementing nonlinear seam elements. The thin-plate formulation has been found to better simulate observed stress distributions in the coal seam and provides a better match to subsidence observations than a simply elastic model [Heasley 1997]. This method is powerful, and it is relatively simple to create the geometric input data.

Some of the limitations of the boundary-element method stem from the basic assumption that the coal seam is surrounded by a homogeneous rock mass consisting of thin, elastic plates separated by zero friction laminations. This method therefore requires that "average" parameters are used for the surrounding rock mass. Since this method assumes the rock mass is elastic, failure and stress redistribution in the surrounding rock are not modeled. These shortcomings can partly be addressed by judicious model calibration against known rock mass response [Heasley 2008].

Full three-dimensional finite-element and finite-difference methods are available that can model the complexity of geometry, geology, and rock failure associated with pillar retreat mining. However, the need to model the rock mass response at a scale of single meters in the vicinity of the pillars while also modeling the surrounding rock mass and mined areas at a scale of more than 1,000 m poses significant challenges in terms of computer resources and run times. A method of simplifying the model geometry while capturing the essential aspects of pillar response was developed by Board and Damjanac [2003] for evaluating the potential for pillar collapse in trona mines. The approach uses equivalent elements that follow the same stress-deformation curve as actual pillars and significantly reduces the need for computer resources while preserving most of the advantages of full three-dimensional models.

METHOD OF EQUIVALENT PILLAR MODELING

Approach

The equivalent pillar modeling method is based on replacing a coal pillar, the surrounding rooms, and the immediate roof and floor by one or more elements that have the same load-deformation response. This allows details of the local pillar, roof, and floor response to be incorporated into larger elements. Using the larger elements, models can be built to include extensive mined areas without sacrificing the effects of local rock mass response. A large-scale model can include any number of different equivalent elements for different room-and-pillar sizes or shapes, allowing typical panel pillars, barriers, and main development pillars to be modeled.

The response of the equivalent elements includes all of the phenomena associated with a pillar undergoing increasing load, such as floor heave, roof collapse, punching of the pillar into weak surrounding strata, and ultimate yield or failure of the pillar (see Figure 1). The detail required to capture each of these events need not be included in the equivalent numerical model, but is implied in the load-deformation response of the equivalent pillar element. After solving a large-scale model of a particular pillar layout, the detailed pillar, roof, or floor response can be found by referring back to the original data or model that was used to develop the load-deformation response.

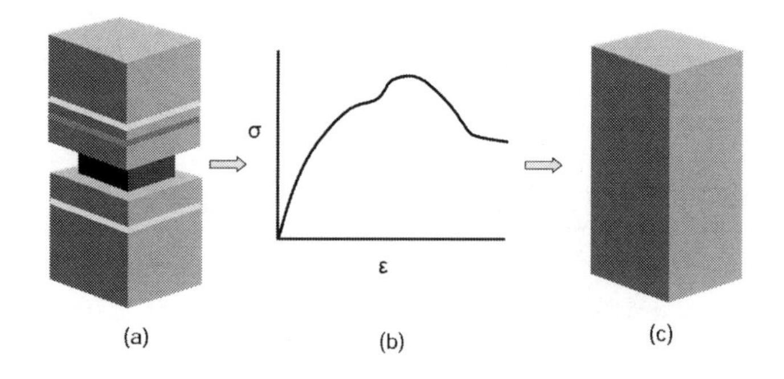

Figure 1.—Concept of an equivalent element model showing (a) the detailed geology and mining geometry being investigated, (b) the resulting stress-strain relationship, and (c) the uniform equivalent element that follows the same stress-strain relationship.

The following discussion is based on the implementation of the equivalent pillar method in the FLAC3D finite-difference program [Itasca Consulting Group 2007]. FLAC3D has an internal programming language called FISH that allows the user to modify the response of individual elements in a model. This facility was used to specify the response of the equivalent pillar elements. The appendix to this paper contains an example of the FISH programming used to define the load-deformation response of a pillar with a width-to-height ratio of 6.0.

Obtaining Pillar Response Curves

The response of a pillar and the immediate roof and floor strata can be obtained from direct measurement and monitoring of pillars in the field. Since appropriate field data are relatively scarce and costly to obtain, numerical models can be used to obtain reasonable estimates of the behavior of pillars under varying geological conditions. These models should be calibrated against field observations where possible.

An example of a FLAC3D numerical model to obtain the load-deformation response of a pillar with a width-to-height ratio of 6.0 is shown in Figure 2. The response of the model pillar to increasing loading was obtained by simulating a downward-moving boundary at the top of the model while fixing the lower boundary and constraining the sides of the model in the horizontal direction. The stress-strain response of the pillar was obtained by recording the average stress at the midheight of the pillar and the strain between points located at the top and bottom of the pillar.

The coal was modeled using the strain-softening Hoek-Brown constitutive model in FLAC3D [Hoek et al. 2002], while the laminated nature of the surrounding rocks was modeled using the bilinear ubiquitous-joint constitutive model. The contact plane between the coal and the surrounding rock was modeled as an explicit interface with a friction angle of 25°. Table 1 summarizes the key material properties used for the coal and the surrounding rock. It was found that the implementation of the Hoek-Brown failure criterion in FLAC3D best represents the performance of coal pillars, especially at width-to-height ratios greater than 4.0.

Table 1.—Properties used to obtain the stress-strain response of a single pillar

Property	Coal	Surrounding rock
Elastic modulus	3 GPa	25 GPa
Uniaxial compressive strength[1]	20 MPa	80 MPa
Hoek-Brown m-parameter	1.47	—
Hoek-Brown s-parameter	0.07	—
Hoek-Brown a-parameter	0.65	—
Friction angle	—	32°

[1]Laboratory-scale strength.

The resulting stress-strain curve for the modeled pillar is shown in Figure 2*B*. It can be seen that the pillar has an initial elastic response up to a stress of about 10 MPa, followed by strain hardening until the pillar reaches a peak strength of 20.4 MPa, followed by strain softening to about 18 MPa. The model predicts a peak strength similar to that predicted by the Mark-Bieniawski equation [Mark 1987], which is widely used in the United States, while the post-

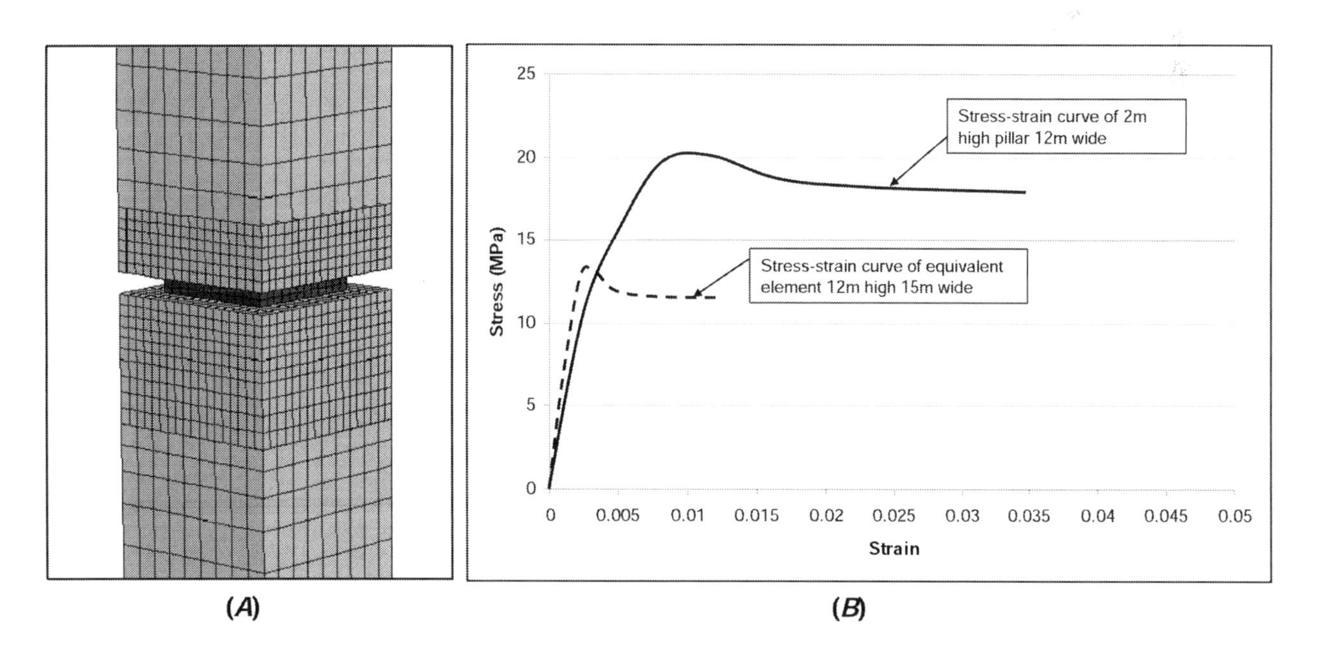

Figure 2.—FLAC3D model of a single pillar with a width-to-height ratio of 6.0, showing the resulting stress-strain curves obtained for the pillar and for the equivalent element. The stress in the equivalent element is lower owing to a larger area of application.

peak strength reduction is similar to field results reported by Mark [1987].

Figure 2*B* also shows the stress-strain curve of an equivalent element that represents the response of the pillar, half of the surrounding entries, and 5 m of the roof and floor rocks. The dimensions of the equivalent element are 15 m by 15 m wide by 12 m high. The curve was obtained by monitoring the average stress across the full width of the model at an elevation 5 m below the midheight of the pillar and monitoring the strain between two points 6 m above and below the midheight of the pillar. The equivalent curve therefore captures the overall behavior of the pillar and the immediately surrounding strata and includes the effects of any local floor heave or roof damage. Both the stress and strain values of the equivalent element seem to be lower than those measured for the pillar because the equivalent element is taller, reducing the strain, and wider, reducing the average stress, than the original coal pillar.

Note that changing the interface properties or the strength of the roof and floor materials can have a significant effect on the peak and residual pillar strength. The stress-strain curve obtained from this analysis can be used to define the properties of equivalent pillar elements in models representing extensively mined and pillared areas.

Modeling the Equivalent Pillars

The key to using equivalent elements in a large-scale model lies in modifying the elements that represent the coal pillars so that they follow the desired stress-strain relationship. We used the Coulomb strain-softening constitutive model in FLAC3D, which can conveniently be modified to achieve the desired stress-strain relationships, after Damjanac [2008]. It is necessary to modify the element behavior so that horizontal confinement will not be generated while it is deformed in the vertical direction, because the effect of confinement is already accounted for in the pillar stress-strain curve. This can be achieved by setting the Poisson's ratio to zero and resetting the horizontal stress components to zero during model solution. Details of the parameter settings and model initiation for FLAC3D are presented in the appendix to this paper.

The equivalent pillar elements used by the authors each simulated the response of a pillar and the surrounding entries up to the center line of the entries. The stress within the equivalent elements will therefore be lower than the stress in the pillar, which has a smaller cross-sectional area. It is therefore necessary to modify the initial stresses in the equivalent elements as follows:

$$\sigma_e = \frac{\sigma_p}{1-r} \qquad (1)$$

where σ_e is the average stress in the equivalent pillar, σ_p is the average stress in the actual pillar, and r is the extraction ratio. When evaluating the results of an analysis, the inverse conversion must be done to obtain the actual pillar stresses from the equivalent stress values reported by the model.

Modeling Abutment Edges and Gob

Crushing of the edges of large abutments or adjoining barrier pillars can also be modeled using equivalent elements. A detailed model of a wide abutment and the adjacent opening can be created, and the average stress-strain response of the outer segment of the abutment can be recorded. Equivalent elements can then be created that follow the same stress-strain relationship. Gob can similarly be modeled by creating equivalent elements that follow the desired stress-strain response.

VERIFICATION OF METHOD

A number of models were created to test the equivalent pillar modeling approach against detailed models of full pillars. Compression testing of single equivalent pillar elements showed that the stress-strain response followed the desired values with an error of less than 1%.

A second test was conducted in which a panel of six entries and five pillars was modeled using both equivalent pillar elements and a detailed pillar model. The pillars were modeled at 200-m depth and the average vertical stress in the pillars was compared. Figure 3 shows the results. Again, it can be seen that the equivalent pillar method provides satisfactory results. The difference between the average stress in the equivalent pillars and the detailed model pillars was less than 2%.

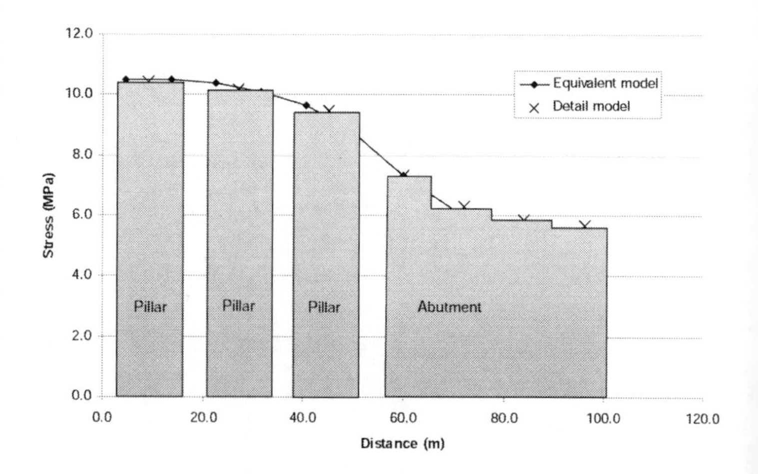

Figure 3.—Pillar and abutment stresses obtained using a detailed model and an equivalent pillar model for a panel of six pillars (only three pillars shown because of symmetry).

EXAMPLE APPLICATION

The example below shows how three-dimensional models have been used to investigate the impact of geology and depth on stress distributions around pillar retreat panels. The results are compared to predictions of the empirically developed ARMPS method.

Mining Geometry

Two models were created—the first simulating retreat mining at a depth of cover of 200 m, the second at 600 m. The geology of the 200-m depth model was selected to simulate a typical extraction panel in the northern Appalachian coalfields with relatively weaker and thinner bedded strata. The model at 600-m depth was set up to simulate pillar extraction in the Western United States, where thicker, stiffer strata are present.

Details of the mining geometry are presented in Table 2. The mining dimensions were selected so that a stable layout would be formed in both cases. Two mining scenarios were considered for each case. The first scenario represented retreat mining in an isolated panel. The second scenario assumed that the active mining panel was located adjacent to a previously mined panel, separated by a barrier pillar.

Table 2.—Mining geometry

Parameter	Value for 200-m depth model	Value for 600-m depth model
Entry and crosscut width	6 m	6 m
Pillar width	18 m	24 m
Mining height	2.4 m	2.4 m
No. of entries in panel	6	5
Width of panel	126 m	126 m
Length of zone containing pillars	300 m	300 m
Length of extracted (gob) zone	300 m	300 m
Barrier width	18 m	60 m
Width of adjacent panel	126 m	126 m

Model Setup

A FLAC3D model was created with horizontal dimensions of 1,100 by 600 m. In the vertical direction, the rock mass was modeled from a point 100 m below the coalbed up to the ground surface. Figures 4 and 5 show examples of the two models, indicating the different degrees of layering of the strata. The side boundaries of the models were constrained in the horizontal direction and the bottom was fixed. The rooms and pillars were modeled using the equivalent pillar approach. The equivalent elements were 12 m high, each representing the 2.4-m-high coal seam and 9.6 m of the surrounding rocks. The abutment edges were also modeled using equivalent elements that represented a 12-m-wide strip of coal with half an entry mined out of it.

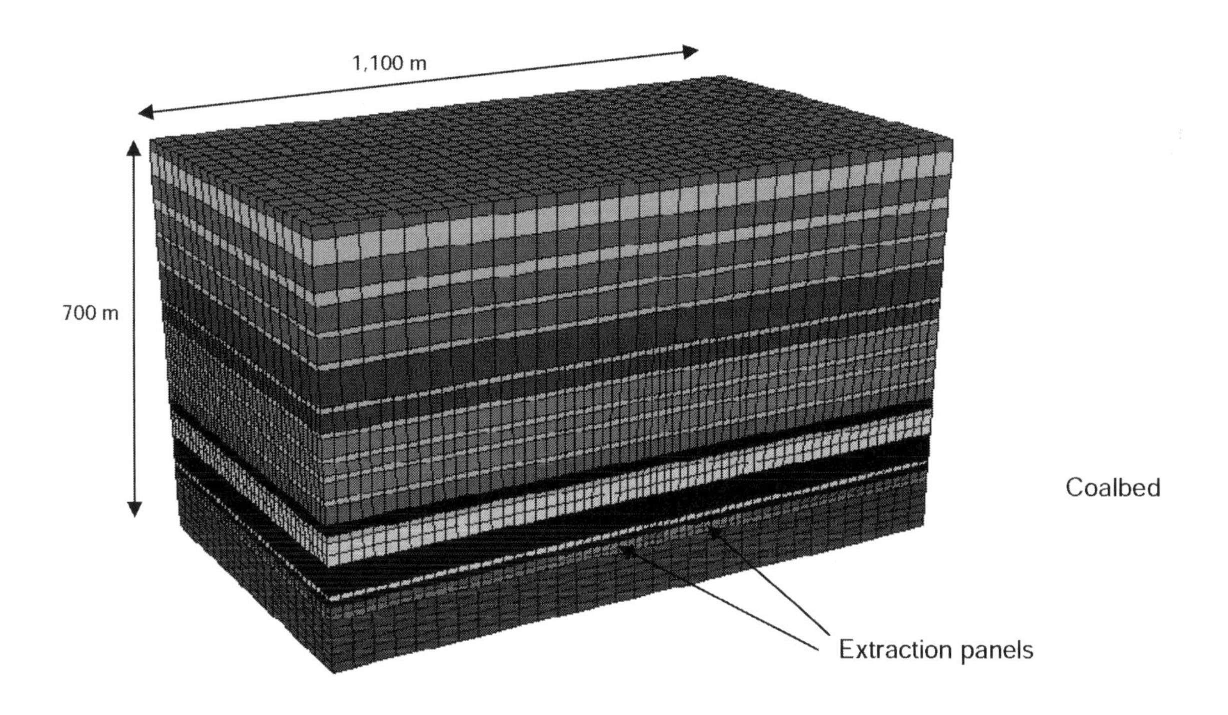

Figure 4.—View of a FLAC3D model of retreat mining at a depth of cover of 600 m in the Western United States using equivalent pillar elements. Rock layering in the model is shown in shades of gray.

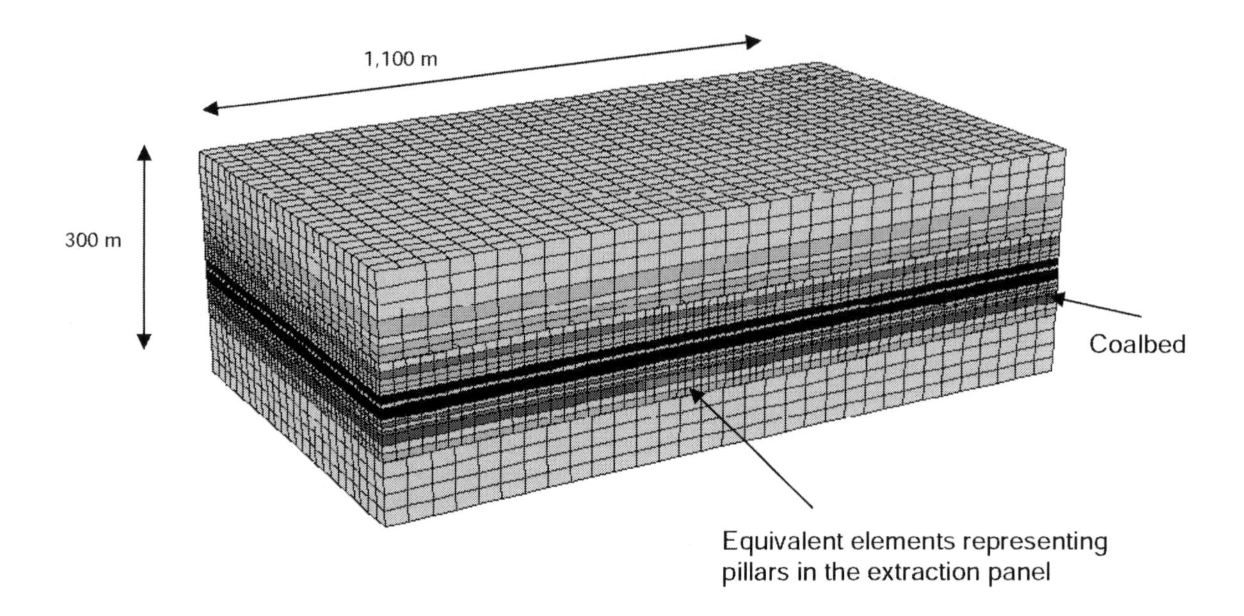

Figure 5.—View of a FLAC3D model of retreat mining at a depth of cover of 200 m in the northern Appalachian coalfields using equivalent pillar elements. Rock layering in the model is shown in shades of gray.

The overburden was modeled using the strain-softening ubiquitous-joint model available in FLAC3D. This model can simulate both bedding plane shearing and intact rock failure, which has been found to be appropriate for modeling the anisotropic strength of bedded coal measure strata [Zipf 2007; Gale 1999]. The rock strength, deformation properties, and bedding strength properties used in the models are presented in Table 3. The compressive strength of the modeled rock was reduced by a factor of 0.58 from the values shown in the table to account for the strength reduction between laboratory samples and the larger-scale in situ rock beds, after Hoek and Brown [1980]. The Poisson's ratio was set at 0.25 and tensile strength set to 10% of the in situ rock compressive strength. Bedding tensile strength was set at 10% of the rock matrix tensile strength. Calibration tests were conducted to ensure that the models provided a reasonable estimate of measured surface subsidence and abutment stresses.

The unmined coalbed was modeled as a solid material using elements that were 12 m high to match the equivalent pillar elements. It was therefore necessary to modify the elastic modulus of the solid coal elements to account for the stiffer rock material included in the 12-m section. The gob was also modeled using the equivalent element approach. It was assumed that complete caving occurred everywhere in the mined area to a height of 6 m above the floor of the coalbed. The stress-strain behavior of the fully caved gob was based on the results of laboratory tests [Pappas and Mark 1993]. For these analyses, the results for strong sandstone were used to model the gob for the deep-cover western example and shale results for the shallower-cover eastern example. Vertical stresses were depth-dependent, and horizontal stresses were initialized using the relationships developed by Mark and Gadde [2008] for U.S. coal mines.

Table 3.—Representative rock properties

Rock type	Uniaxial compressive strength (MPa)	Elastic modulus (GPa)	Rock matrix friction angle (degrees)	Bedding cohesion (MPa)	Bedding friction angle (degrees)
Weak shale	20	7.0	20	0.3	5
Moderate-strength shale	40	11.0	25	0.5	7
Moderate-strength sandstone	60	15.0	35	2.0	10
Strong sandstone	100	20.0	40	4.0	12
Moderate-strength limestone	80	30.0	40	6.0	15
Strong limestone	140	40.0	42	12.0	25

Table 4.—Results obtained by FLAC model and ARMPS empirical method

Parameter	Eastern U.S. model			Western U.S. model		
	FLAC	ARMPS	Difference (ARMPS: FLAC)	FLAC	ARMPS	Difference (ARMPS: FLAC)
Average pillar stress after development (MPa)	8.2	9.1	+11.0%	22.2	24.2	+9.2%
Average stress in AMZ first panel mining (MPa)	13.3	13.8	+3.8%	26.0	33.2	+28.0%
Average stress in AMZ second panel mining (MPa)	16.7	15.4	−7.7%	28.9	34.5	+19.0%

After building each model, the vertical stresses within the equivalent elements were modified to account for the presence of the excavated rooms using Equation 1. The model was then allowed to equilibrate with these modified stresses. Mining was then modeled in a stepwise manner, extracting the equivalent pillar elements and replacing them with gob elements until the length of the extracted zone reached 300 m. At this stage, the gob zone was 300 m long and the zone containing the remaining pillars was also 300 m long. Owing to symmetry conditions, the effective length of these zones was twice the modeled length. The model was run to equilibrium at each step, and the results of the final step are discussed below.

RESULTS

Table 4 compares the results obtained by the FLAC3D model using the equivalent element approach with the empirically developed ARMPS results within the active mining zone (AMZ). The AMZ represents a group of pillars at the extraction front that are assumed to carry 90% of the front abutment load [Mark and Chase 1997]. The results presented here are applicable only for the particular geometry and geology modeled and should not be considered to be generally valid for other mine geometries or geological conditions.

The results show that at the development stage ARMPS predicts pillar stresses that are about 10% higher than the FLAC predictions. The main reason for this difference is that ARMPS assumes that the development pillars are carrying the full tributary loading, while FLAC considers the reduced stiffness of the pillars, allowing some of this load to be distributed to other areas.

The stresses in the AMZ seem to be highly dependent on the stiffness ratio of the overlying strata to the pillar stiffness. The results show that for the western U.S. model with stiffer overburden, ARMPS predicts an average AMZ stress that is 28% higher than that calculated by FLAC. However, the results for the eastern U.S. case differ by only 3.8%. This difference is significant and can partially explain why deep-cover retreat mining has been successful at relatively low values of the ARMPS stability factor. The results also show that the three-dimensional models that include details of the geology, relative rock stiffness, pillar response, horizontal stress, gob, and rock mass failure can provide additional insight into the complex rock response resulting from retreat mining.

CONCLUSIONS

The method of equivalent pillar modeling allows three-dimensional stress analysis of large arrays of pillars to be carried out efficiently without losing the essential aspects of the interaction between the pillars and the surrounding rocks. The method allows models of large areas to be created that are manageable in terms of computing requirements. Testing of the technique through comparison to a detailed three-dimensional model shows that the equivalent models accurately represent the overall response of the rock mass and load distribution among the pillars and surrounding unmined coal.

An example application has shown that the equivalent pillar technique applied to three-dimensional models can improve our understanding of the interaction between surrounding strata and pillars during retreat mining. The impact of variations in geology and geometry can readily be assessed. The developed method can therefore contribute to improved design guidelines and greater safety in retreat mining.

REFERENCES

Board M, Damjanac B [2003]. Development of a methodology for analysis of instability in room and pillar mines. In: Stephansson O, ed. Proceedings of the 2003 Swedish Rock Mechanics Day Conference. Stockholm: SveBeFo, pp. 1–22.

Chase FE, Mark C, Heasley KA [2002]. Deep-cover pillar extraction in the U.S. coalfields. In: Peng SS, Mark C, Khair AW, Heasley KA, eds. Proceedings of the 21st International Conference on Ground Control in Mining. Morgantown, WV: West Virginia University, pp. 68–80.

Damjanac B [2008]. Personal communication. Minneapolis, MN: Itasca Consulting Group, Inc., April 2008.

Gale WJ [1999]. Experience of field measurement and computer simulation methods of pillar design. In: Mark C, Heasley KA, Iannacchione AT, Tuchman RJ, eds. Proceedings of the Second International Workshop on Coal Pillar Mechanics and Design. Pittsburgh, PA: U.S. Department of Health and Human Services, Centers for Disease Control and Prevention, National Institute for Occupational Safety and Health, DHHS (NIOSH) Publication No. 99–114, IC 9448, pp. 49–61.

Gates RA, Gauna M, Morley TA, O'Donnell JR Jr., Smith GE, Watkins TR, Weaver CA, Zelanko JC [2008].

Report of investigation: underground coal mine, fatal underground coal burst accidents, August 6 and 16, 2007, Crandall Canyon mine, Genwal Resources, Inc., Huntington, Emery County, Utah, ID No. 42-01715. Arlington, VA: U.S. Department of Labor, Mine Safety and Health Administration.

Heasley KA [1997]. A new laminated overburden model for coal mine design. In: Mark C, Tuchman RJ, eds. Proceedings: New technology for ground control in retreat mining. Pittsburgh, PA: U.S. Department of Health and Human Services, Centers for Disease Control and Prevention, National Institute for Occupational Safety and Health, IC 9446, pp. 60–73.

Heasley KA [2008]. Some thoughts on calibrating LaModel. In: Peng SS, Tadolini SC, Mark C, Finfinger GL, Heasley KA, Khair AW, Luo Y, eds. Proceedings of the 27th International Conference on Ground Control in Mining. Morgantown, WV: West Virginia University, pp. 7–13.

Hoek E, Brown ET [1980]. Underground excavations in rock. London: Institution of Mining and Metallurgy.

Hoek E, Carranza-Torres C, Corkum B [2002]. Hoek-Brown failure criterion: 2002 edition. In: Proceedings of the Fifth North American Rock Mechanics Symposium (Toronto, Ontario, Canada), Vol. 1, pp. 267–273.

Itasca Consulting Group [2007]. Fast Lagrangian analysis of continua (FLAC3D). Minneapolis, MN: Itasca Consulting Group, Inc.

Mark C [1987]. Analysis of longwall pillar stability [Dissertation]. University Park, PA: The Pennsylvania State University, Department of Mining Engineering.

Mark C, Chase FE [1997]. Analysis of retreat mining pillar stability (ARMPS). In: Mark C, Tuchman RJ, eds. Proceedings: New technology for ground control in retreat mining. Pittsburgh, PA: U.S. Department of Health and Human Services, Centers for Disease Control and Prevention, National Institute for Occupational Safety and Health, IC 9446, pp. 17–34.

Mark C, Gadde M [2008]. Global trends in coal mine horizontal stress measurements. In: Peng SS, Tadolini SC, Mark C, Finfinger GL, Heasley KA, Khair AW, Luo Y, eds. Proceedings of the 27th International Conference on Ground Control in Mining. Morgantown, WV: West Virginia University, pp. 319–331.

Mark C, Zelanko JC [2005]. Reducing roof fall accidents on retreat mining sections. Coal Age *110*(12):26–31.

Mark C, Chase FE, Pappas DM [2003]. Reducing the risk of ground falls during pillar recovery. In: Yernberg WR, ed. Transactions of the Society for Mining, Metallurgy, and Exploration, Inc. Vol. 314. Littleton, CO: Society for Mining, Metallurgy, and Exploration, Inc., pp. 153–160.

Mark C, Chase FE, Pappas DM [2007]. Analysis of multiple-seam stability. In: Peng SS, Mark C, Finfinger GL, Tadolini SC, Khair AW, Heasley KA, Luo Y, eds. Proceedings of the 26th International Conference on Ground Control in Mining. Morgantown, WV: West Virginia University, pp. 5–18.

Pappas DM, Mark C [1993]. Behavior of simulated longwall gob material. Pittsburgh, PA: U.S. Department of the Interior, Bureau of Mines, RI 9458. NTIS No. PB93-198034.

Salamon MDG [1991]. Deformation of stratified rock masses: a laminated model. J S Afr Inst Min Metall *91*(1): 9–26.

Skiles ME, Stricklin KG [2008]. Technical support assistance in reviewing roof control plans. Arlington, VA: Mine Safety and Health Administration, Procedure Instruction Letter No. I08-V-02, May 29, 2008. Available at: http://www.msha.gov/regs/complian/pils/2008/pil08-v-2.pdf

Stricklin KG, Skiles ME [2008]. Precautions for the use of the Analysis of Retreat Mining Pillar Stability (ARMPS) computer program. Arlington, VA: Mine Safety and Health Administration, Program Information Bulletin No. P08-08, April 7, 2008. Available at: http://www.msha.gov/regs/complian/PIB/2008/pib08-08.asp

Zipf RK Jr. [2007]. Numerical modeling procedures for practical coal mine design. In: Mark C, Pakalnis R, Tuchman RJ, eds. Proceedings of the International Workshop on Rock Mass Classification in Underground Mining. Pittsburgh, PA: U.S. Department of Health and Human Services, Centers for Disease Control and Prevention, National Institute for Occupational Safety and Health, DHHS (NIOSH) Publication No. 2007–128, IC 9498, pp. 153–162.

APPENDIX.—CREATING EQUIVALENT PILLAR ELEMENTS IN FLAC3D

Equivalent pillar elements respond to the vertical closure between the roof and floor in the same way that a pillar would. In FLAC3D, this can be achieved by using the Coulomb strain-softening logic to control the response of the equivalent pillar elements.

When using the strain-softening approach, a number of modifications to the element properties must be made so that they perform as required. The Poisson's ratio of the elements must be set to zero so that lateral dilation does not occur, which can cause additional lateral confinement to the equivalent pillar elements. For convenience, the friction angle is also set to zero so that the strength and stress-strain response of the equivalent pillar element can be controlled by varying the cohesion, described below.

The equivalent elements will respond elastically until their strength is exceeded. The evolution of the pillar strength after the initial elastic response, through strain hardening, strain softening, and ultimate plastic yield, is specified by a FLAC3D table, which relates the element cohesion to the plastic strain. Since the friction angle is set to zero, the cohesion should be one-half the desired strength. Table A-1 shows the calculation of the cohesion-strain values used to define the response of an equivalent element that simulates the 6.0 width-to-height ratio shown in Figure 2. This table assumes the equivalent pillar responds elastically up to 6.4 MPa, then departs from elastic behavior up to the peak stress of 12.8 MPa, followed by yielding to a residual strength of 11.52 MPa. At each point on the stress-strain curve, the elastic component of strain is subtracted so that only the plastic component of strain is used when generating the strain-cohesion pairs. An elastic modulus of 7.2 GPa was used to calculate the elastic strains. This value represents the combined modulus of the rock and coal in the equivalent element and can be obtained from the stress-strain curve of the equivalent element or approximated by calculation.

Table A-1.—Calculation of plastic strain-cohesion values to control equivalent element yielding

Equivalent stress (MPa)	Total strain	Elastic strain	Plastic strain	Cohesion
0	0.0000	0.0000	0.0000	0
6.40	0.0009	0.0009	0.0000	3.20
10.88	0.0020	0.0015	0.0005	5.44
12.80	0.0035	0.0018	0.0017	6.40
11.52	0.0067	0.0016	0.0051	5.76
11.52	0.0160	0.0016	0.0144	5.76

The following text shows how the strain-softening properties of the equivalent elements are defined in FLAC, followed by the FISH programming to implement the equivalent element behavior. In this example, the pillars were created with a group name "P1". The pillars are first defined as strain-softening, and the stress-strain relationship for the pillars is provided as a cohesion table. The "countpillar" function counts the number of pillars of type "P1" in the model and sets up an array to store the memory addresses of these pillars. The "parray" function searched for the pillars of type "P1" and stores their memory addresses in the aforementioned array. It also calculates and sets the initial stress in the equivalent pillar elements. The "dopillar" function destroys the horizontal stress in the equivalent pillar elements during each solution cycle.

```
{Assign pillar properties in command mode}
{Cohesion is one-half desired pillar strength}

 model ss range group P1
 prop dens=2000 b=bmod s=smod  range   group P1
 prop fric=0 coh=3.2e6 ctable=10 range group P1

 {Create cohesion table - half of desired strength –
obtained from Table A-1 above}

 Table 10 0,3.20e+06 0.0005,5.44e+06 0.0018,6.40e+06
0.0051,5.76e+06 0.0144,5.76e+06

{Count the number of pillar elements in group P1}

def countpillar
  pnt = zone_head
  npillar1 = 0
  loop while pnt # null
    if z_model(pnt) # 'null'
      if z_group(pnt) = 'P1'
        npillar1 = npillar1 + 1
      endif
    endif ;not null element
  pnt = z_next(pnt)
  endloop

{Create an array to store pointers to the elements in
group P1}

 Parraysize1 = npillar1
 if Parraysize1 = 0 then
  Parraysize1 = 1
  endif
end ;countpillar
countpillar
```

{Populate the array with pointers to the P1 elements}

```
def parray
if npillar1 > 0
array pelts1(Parraysize1)
  i=1
  pnt = zone_head
  loop while pnt # null
    if z_model(pnt) # 'null'
      if z_group(pnt) = 'P1'
        pelts1(i) = pnt
        i = i + 1
      endif ;belongs to group P1
    endif  ;not null element
  pnt = z_next(pnt)
  endloop
```

{Fix initial stress in equivalent pillar elements based on extraction}

```
percpillar1 = 0.61
; percpillar is percentage pillars = 1-extraction ratio
  loop iz (1,npillar1)
    z_szz(pelts1(iz))=z_szz(pelts1(iz)) * percpillar1
    z_extra(pelts1(iz),1) = percpillar1
  endloop
endif ; if npillar1
end
parray
```

{Routine to destroy horizontal stress in equivalent pillars P1}

```
def dopillar
whilestepping
if npillar1 > 0
loop iz (1,npillar1)
   z_sxx(pelts1(iz))=0.0
   z_syy(pelts1(iz))=0.0
   z_sxy(pelts1(iz))=0.0
   z_sxz(pelts1(iz))=0.0
   z_syz(pelts1(iz))=0.0
endloop
endif
end
```

NUMERICAL MODEL EVALUATION OF FLOOR-BEARING CAPACITY IN COAL MINES

By Murali M. Gadde[1]

ABSTRACT

This paper addresses the question of ultimate floor-bearing capacity for pillar design in room-and-pillar coal mines. It lists some of the idiosyncrasies that limit the application of classical bearing capacity analyses that are based on closed-form solutions. An approach that uses numerical models is described. The models are first validated against a simple case of a strip footing resting on a semi-infinite homogeneous weightless soil. Two cases are then presented in which calibrated three-dimensional models are used to investigate the effect of nonuniform distribution of pillar loads and the effect of interaction between pillars on floor-bearing capacity. The results show that nonuniform pillar loading of the floor is not likely to cause a significant change in ultimate bearing capacity of the floor, but can explain superficial floor heave that is often observed in coal mines. The interaction between adjacent pillars can have a significant effect on pillar strength if the weak floor has a non-zero friction angle. Results show that the strength of the floor adjacent to rib pillars experiences greater benefit from interaction compared to rectangular pillars. It is concluded that the design of coal mine workings can be adequately conducted by using numerical models that simulate the complex pillar-floor interactions.

INTRODUCTION

Coal mines in the Illinois basin extract coal using the room-and-pillar method. The coalbeds are often underlain by weak strata that can fail when the pillar load exceeds the bearing capacity, resulting in floor heave and excessive closure in the mine workings that can cause surface subsidence. Classical bearing capacity models do not account for the idiosyncrasies of a typical coal mine floor stability problem, shown in Figure 1, including variations of pillar length (L) and width (B), entry width (s), thickness of weak floor bed (H), and variations in strata cohesion (c), friction angle (φ), and density (γ). A determination of the floor-bearing capacity in a coal mine should consider the following:

- Different geometries of pillars in the plan view (square, rectangular, long-continuous, parallelogram, and irregular)
- Multiple layers of strata with variable thickness could exist in the floor within the zone of influence of a coal pillar
- Presence of multiple pillars in close proximity at uniform or variable spacing
- Each layer of rock in the floor is deformable and normally has non-zero values of cohesion, friction angle, and density
- Some of the floor strata may not be adequately described by the Mohr-Coulomb failure criterion
- Volumetric expansion of the floor strata is possible in the postfailure state (dilatation effects may not be ignored)
- Spatial variation of floor properties (both laterally and with depth)
- Time-dependent deformation and failure of the floor
- Effect of any water accumulation on time-dependent pore water pressure changes and accompanying strength degradation
- Nonuniform vertical stress distribution on coal pillars
- Presence of in situ horizontal stresses

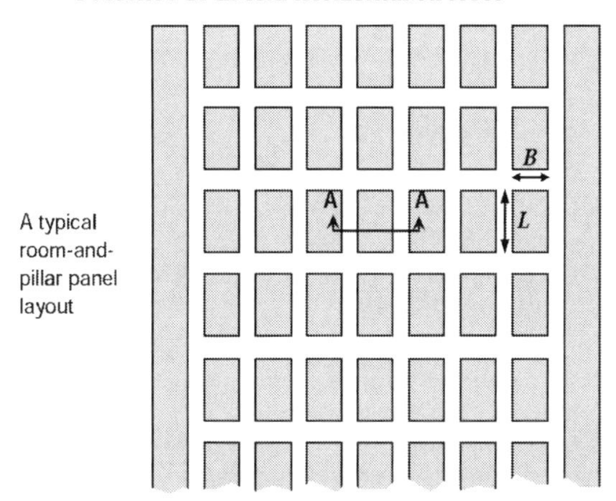

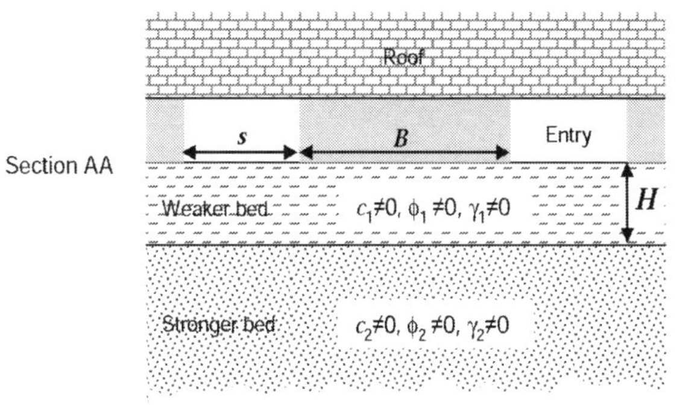

Figure 1.—Features of a typical coal mine floor-bearing capacity problem.

[1]Peabody Energy, St. Louis, MO.

Conventional bearing capacity solutions treat the floor as an isolated component of the underground opening stability. In reality, however, the floor, coal pillar, and roof function together as a system and interact with each other. It is often necessary to consider the entire system stability rather than any one component's integrity. In such cases, the analytical method should be able to incorporate the interaction effects of the different components of the entire system.

When all of the requirements set forth are considered, it is impractical to provide a closed-form solution to the coal mine bearing capacity problem. There is, however, an alternative. With the tremendous advancements made in computer hardware and the advent of sophisticated numerical modeling tools over the past 2 decades, it is now possible to incorporate almost every single aspect of the bearing capacity problem mentioned above in the analysis. Such flexibility and versatility is achieved because numerical models deal with a complex problem by following a "parts-to-whole" approach. In a numerical model, the final solution to a "big" problem is accomplished by putting together solutions to several "small" problems. It is easier to incorporate any degree of complexity at the "parts" level rather than at the "whole" level as accomplished in a closed-form solution.

Use of numerical modeling to investigate coal mine floor stability is not new. Some interesting work was done in the past by researchers from several coal-producing countries [Rockaway and Stephenson 1979; Chandrashekhar 1990; Bandopadhay 1982; Deb et al. 2000; Vasundhara 1999; Bhattacharyya and Seneviratne 1992; Yavuz et al. 2003]. The past studies, however, relied heavily on two-dimensional modeling to explain some site-specific field behavior or were limited to a few parametric studies of limited applicability.

This paper is devoted to applying numerical modeling to study two aspects of the floor stability problem: (1) the interaction effect of multiple pillars and (2) nonuniform distribution of pillar loads on floor-bearing capacity. The studies were conducted using the FLAC3D numerical modeling code developed by the Itasca Consulting Group, Inc. The explicit Lagrangian solution scheme and the mixed discretization procedures adopted in FLAC3D using the finite-difference approach makes it a powerful tool to address nonlinear problems. Further, the explicit scheme makes it easier to apply loads and deformations in a manner analogous to physical tests. Such a solution methodology also provides the model response in a physically comparable manner. FLAC3D provides a programming language called FISH, which facilitates addition of functionalities that are not included in the standard features of the program and helps automate several aspects of modeling.

MODEL VALIDATION

The first step in developing a proper numerical modeling methodology is to "calibrate" the process such that theoretical solutions are reproduced within a reasonable tolerance. This calibration is necessary because the modeling results are sensitive to the mesh size and loading rates adopted for the simulation. This particular mesh density and the loading rate value, which provides a match with the theoretical solutions, can then be used for further studies.

The simplest bearing capacity problem for which an exact closed-form solution exists is that of a strip footing resting on a semi-infinite homogeneous weightless soil, known as Prandtl's solution [Terzaghi and Peck 1967]. This situation was modeled to verify the modeling methodology. For the verification models, the cohesion value was arbitrarily chosen as 150 psi and the friction angle was varied from 0° to 30° in 5° increments. A constant Poisson's ratio equal to 0.35 was used in all of the models. The soil was assumed to satisfy the Mohr-Coulomb yield function with perfectly plastic behavior. It was also assumed that the soil did not exhibit volume change in the postyield state. Although a non-zero density value was input for the floor in the models, the computed bearing capacity would almost be the same as that of a weightless soil because the width of the strip load was only 12 in. The mesh density and the loading rate were calibrated for the 5° friction angle case and then kept constant for much of the research. Of course, it was not possible to get an exact match for the mesh density when different geometric parameters are varied, but every effort was made to keep the discretization as close as possible to the one used in the verification models.

The modeling results indicated that the verification models will provide bearing capacity results that match to within 3.5% of the theoretical values for a range of material friction angles. The computed and theoretical bearing capacity factors (N) are presented in Table 1, which shows the results for both a smooth and rough footing. Clearly, the numerical models simulate the bearing capacity problem with sufficient accuracy. The validated numerical modeling approach, model discretization, and loading rates were used to conduct the further studies presented in this paper.

Table 1.—Comparison of theoretical- and modeling-based bearing capacity factor, N_c, for a constant cohesion equal to 150 psi

Friction angle, degrees	Bearing capacity factor, N_c		
	Theoretical	Model: smooth footing	Model: rough footing
0	5.14	5.15	5.24
5	6.49	6.49	6.65
10	8.35	8.31	8.61
15	10.99	10.99	11.36
20	14.85	14.89	15.37
25	20.76	20.80	21.47
30	30.22	30.23	30.52

EVALUATION OF NONUNIFORM PILLAR STRESS DISTRIBUTION

In all of the traditional bearing capacity theories, the vertical load on the footing is assumed to be uniformly distributed. However, in underground coal mines, it is known that the vertical stress over a pillar is not uniform. Although some researchers [Chugh and Pytel 1992] have recognized this issue before, no study has ever been done on the effect of nonuniform stress distributions on the computed bearing capacity. The numerical modeling methodology developed here offers an opportunity to study this aspect of the bearing capacity problem.

When a realistic coal mine geometry is considered, in addition to the nonuniform stress distribution over the coal pillar, some horizontal premining stresses also exist in the floor. For the "soft" underclay material, the magnitude of the horizontal stresses may not be very high. Nevertheless, some non-zero horizontal stresses exist in the floor. To account for all of these effects, an actual case of a coal mine with weak immediate floor has been chosen. For the sake of simplicity, only two-dimensional modeling has been conducted here. The geometry near the coal seam used in the model is shown in Figure 2. The model contains an 18-ft-wide entry and 52-ft-wide pillar resting on a 7-ft-thick claystone floor.

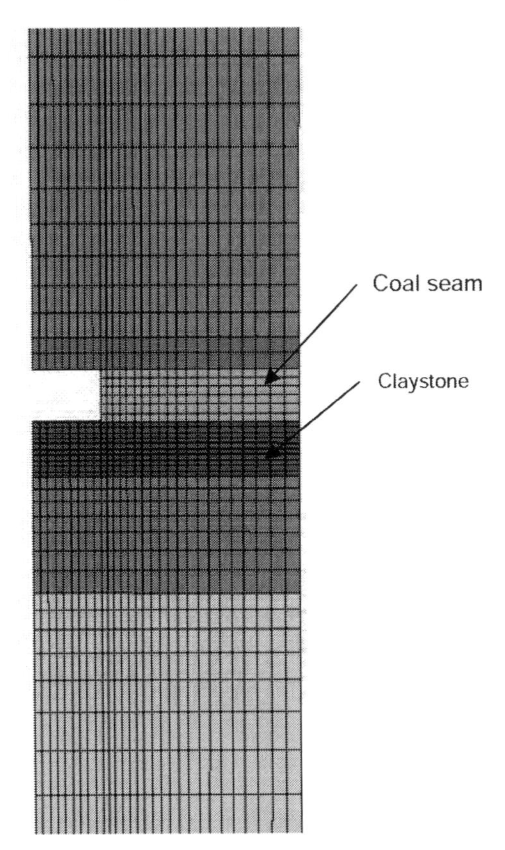

Figure 2.—Part of the modeled geometry to study the nonuniform vertical stress distribution effect on bearing capacity.

In the above model, the claystone floor has been assumed to behave in a perfectly plastic manner, satisfying the Mohr-Coulomb yield criterion. Arbitrarily chosen cohesion and friction angle values equal to 300 psi and 20°, respectively, were assigned to the claystone floor. The rest of the model was assigned elastic properties. After the initial model was solved to incorporate Poisson's ratio-based horizontal stresses, the mine opening was created and the model was solved again to equilibrium. At this stage, the top boundary of the model was fixed in the vertical direction and displacement-controlled load was imposed on the coal pillar until the floor material completely reached its limit state. The resultant vertical stress on the first layer of floor below the pillar was monitored continuously as the model was being solved using a FISH function. The simulation is shown in Figure 3a, where the resultant vertical stress on the floor below the coal pillar is plotted against the number of model steps.

Similarly, to simulate a rigid plate loading on the floor in a different model, the entire cover above the coal pillar was removed and the displacement loading was applied over the top of the coal pillar. This later model was not solved for equilibrium before commencing the displacement loading. The vertical stress variation on the floor for this loading situation is plotted in Figure 3b.

The modeling results showed that the limit stress value was 828 psi for the nonuniform pillar loading case and 800 psi for the rigid plate condition for the assumed inputs. Therefore, consideration of the realistic vertical stress distribution on the coal pillar did not alter the ultimate bearing capacity by more than about 3%. The slightly higher bearing capacity in the nonuniform load case was due to the existence of small horizontal stresses in the floor compared to the uniform loading case, where no horizontal stresses were applied. At the limit state, the extent of yielding and the shear strain rate distribution also looked similar in both cases. There is, however, one major difference. As shown by the red ellipses in Figure 3a, several localized floor failures occurred before reaching the limit state. For this case, it was found that when additional loads were induced in the coal pillar by the uniform displacement loading, in areas close to the pillar rib where the highest vertical stress was acting before, floor failure commenced. As the floor below the rib failed, the vertical stress was transferred deeper into the coal pillar, thus establishing another stress concentration zone. With further displacement loading, progressively deeper portions of the floor reached limit states in a rather abrupt fashion, resulting in the stairstepped stress variation shown in Figure 3a. Such stress variation, however, does not seem to have altered the ultimate bearing capacity of the floor.

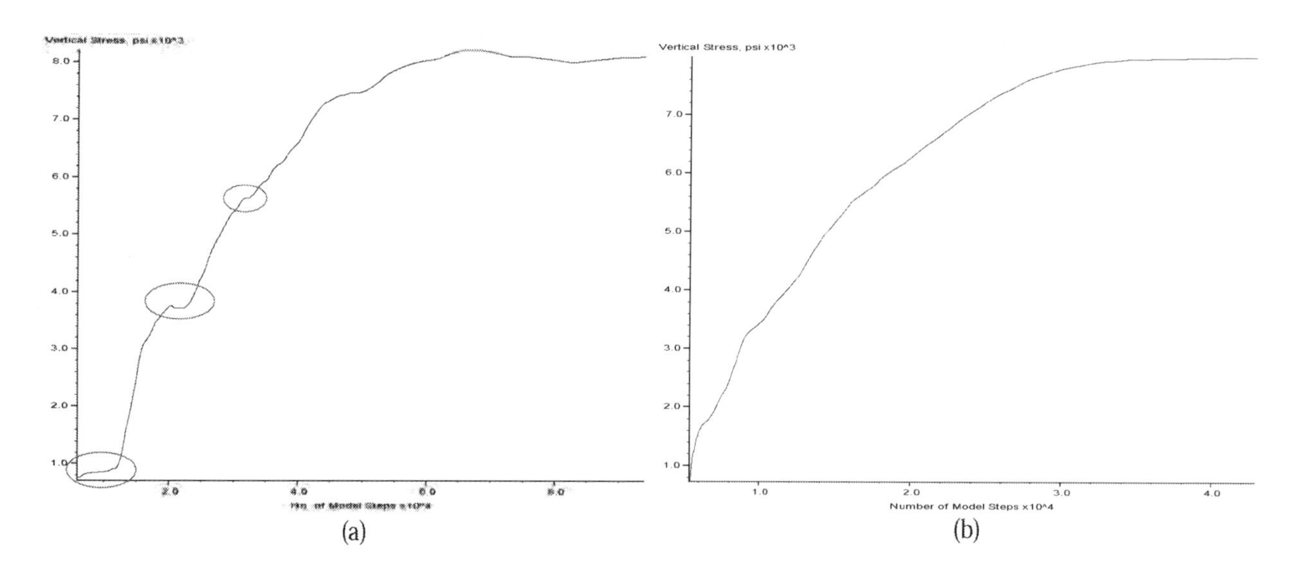

Figure 3.—The variation of resultant vertical stress on floor for (a) nonuniform pillar stress distribution and (b) uniform pillar stress distribution.

It must be reminded here that under a rigid plate or when uniform displacement loading is applied to the floor, the vertical stress on the floor is nonuniformly distributed. Therefore, progressive failure of floor occurs even without mining-induced vertical stresses when the loading is applied through uniform vertical displacement. However, when mining-induced vertical stress acts on the pillar, the extent of nonuniformity becomes much larger. Besides, applying the uniform vertical displacement at the top of the deformable coal pillar reduces the amount of nonuniformity in vertical stress distribution as opposed to a rigid plate situation.

The localized floor failures due to the nonuniform vertical stress explain the practical observation that every floor instability seen underground (Figure 4) does not lead to surface subsidence. Under the influence of nonuniform vertical pillar stress, areas close to the pillar rib could experience localized floor failure, resulting in heave in the adjacent mine opening, while the larger portion of the pillar itself has stable floor. Detailed in situ investigations by Vasundhara et al. [2001] in Australia prove the validity of this conclusion. Therefore, from a subsidence prevention point of view, it is the limit state that matters, not the localized floor instabilities such as that shown in Figure 4. This observation indicates that when collecting data to develop design guidelines for surface subsidence prevention, localized underground instabilities should not be considered unless there is evidence that the floor failures led to surface movements.

Although very limited study was done, the preliminary results given here show that when computing the ultimate bearing capacity, the difference between the uniform displacement loading and nonuniform stress distribution is negligible. The nature of stress distribution seems to affect only the path traversed to reach the limit state, but not the limit value itself. Therefore, the results obtained under uniform displacement loading could perhaps be used without incurring significant errors in the computed bearing capacity.

Figure 4.—Localized floor instability in an Illinois basin coal mine.

PILLAR SPACING EFFECT

In every coal mine, multiple pillars exist in close proximity to each other. Thus, the influence of interference effects on floor-bearing capacity cannot be ignored. For the sake of brevity, the limits of bearing capacity were determined for extreme values for the model inputs. Such a study will show the maximum possible influence of the presence of multiple pillars on the computed floor-bearing capacity for typical Illinois basin coal mine conditions.

The following extremes for different model inputs will include most of the Illinois basin coal mines:

Pillar geometry	*Strip pillar*	*Square pillar*
B/H	40	4
s/B	0.2	0.66
c_2/c_1	2 psi	5 psi
ϕ_1, ϕ_2	0°	35°

For all of the models, symmetry conditions were exploited to simulate the effect of the presence of multiple pillars. The floor materials were assumed to behave in a perfectly plastic manner, satisfying the Mohr-Coulomb yield criterion. Each layer of floor by itself was assumed to be homogeneous and isotropic. Uniform loading on the floor was applied through displacements at the top of the floor over the region represented by the pillar. The floor contact conditions were assumed to be smooth. Typical meshes used for one case each of strip and square pillar are shown in Figure 5. The model results are presented in Tables 2 and 3 for the square and strip footing cases, respectively. The influence of multiple pillar interaction is assessed by the ratio of model strengths computed with and without the presence of multiple pillars. The ratio will show the exact extent of the interaction effect and the potential error when using traditional bearing capacity equations that ignore the effect of adjacent pillars.

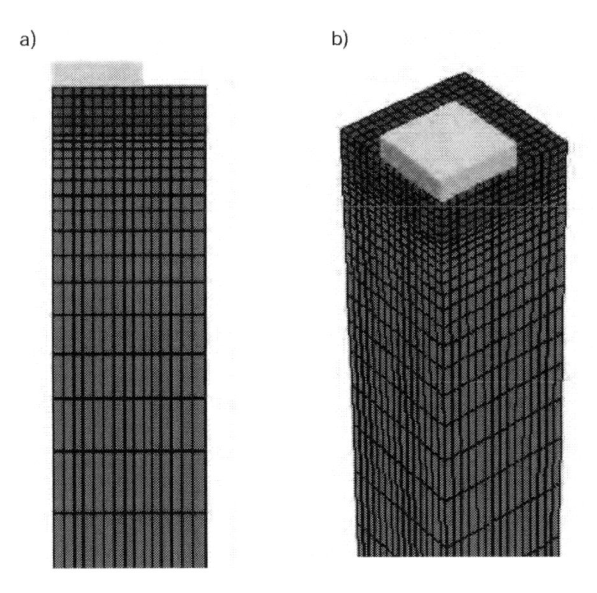

Figure 5.—Typical meshes used for the simulation of adjacent pillar effect for (a) strip and (b) square pillars.

Table 2.—Effect of interference of multiple square footings in close proximity

ϕ_1	ϕ_2	c_2/c_1	B/H	s/B	Ratio of model strength with and without adjacent pillars
0	0	2	40	0.66	1.00
0	0	2	40	0.20	1.05
0	35	2	40	0.66	0.98
0	35	2	40	0.20	1.03
35	35	2	40	0.66	1.22
35	35	2	40	0.20	>4.25
0	0	2	4	0.66	1.00
0	0	2	4	0.20	1.05
0	35	2	4	0.66	1.00
0	35	2	4	0.20	1.05
35	35	2	4	0.66	1.05
35	35	2	4	0.20	>3.4
0	0	5	40	0.66	1.00
0	0	5	40	0.20	1.03
0	35	5	40	0.66	0.98
0	35	5	40	0.20	0.99
35	35	5	40	0.66	1.31
35	35	5	40	0.20	>2.6
0	0	5	4	0.66	1.00
0	0	5	4	0.20	1.05
0	35	5	4	0.66	1.00
0	35	5	4	0.20	1.04
35	35	5	4	0.66	0.94
35	35	5	4	0.20	>2

Table 3.—Effect of interference of multiple strip footings in close proximity

ϕ_1	ϕ_2	c_2/c_1	B/H	s/B	Ratio of model strength with and without adjacent pillars
0	0	2	40	0.66	1.02
0	0	2	40	0.20	1.18
0	35	2	40	0.66	1.01
0	35	2	40	0.20	1.01
35	35	2	40	0.66	2.60
35	35	2	40	0.20	>8
0	0	2	4	0.66	1.19
0	0	2	4	0.20	1.26
0	35	2	4	0.66	1.19
0	35	2	4	0.20	1.26
35	35	2	4	0.66	2.35
35	35	2	4	0.20	>6
0	0	5	40	0.66	1.02
0	0	5	40	0.20	1.04
0	35	5	40	0.66	1.01
0	35	5	40	0.20	1.00
35	35	5	40	0.66	2.61
35	35	5	40	0.20	>3.5
0	0	5	4	0.66	1.19
0	0	5	4	0.20	1.26
0	35	5	4	0.66	1.19
0	35	5	4	0.20	1.26
35	35	5	4	0.66	1.80
35	35	5	4	0.20	>3.6

The major conclusions from the models can be summarized as follows:

- When the friction angle of both floor layers is zero, the interaction effect of adjacent pillars is negligible.
- Irrespective of the *B/H* value, if the top layer is frictionless even when the stronger layer has non-zero friction angle value, there is negligible interaction effect.
- When the friction angle of the weak floor layer is zero and the ratio of *B/H* is small, there is negligible influence of the adjacent pillar.
- Interaction of adjacent pillars will influence the bearing capacity ***only*** if the immediate floor has non-zero angle of internal friction. When both layers have $\phi = 35°$ and $s/B = 0.2$, the models did not reach limit state even after solving for more than 500,000 to 1 million model steps. The ratios for the strength values in Tables 2 and 3 are based on the resultant floor stress obtained when the solution was terminated. Therefore, if both the layers of floor beds have higher non-zero friction values, then such conditions result in virtually indestructible floor, and the pillar design may depend on the coal strength rather than the floor strength.
- In almost every case studied, the effect of adjacent pillars on bearing capacity is higher for strip pillars than square pillars. Therefore, rectangular pillars will gain more from their neighbors than square pillars of the same width.

The reason for the dramatic increase in the bearing capacity with the friction angle in the presence of multiple pillars could be seen from the minimum principal stress contours and the displacement vectors plotted in Figure 6. As the vertical load on the pillar increased, the horizontal floor movements induced in the adjacent entry were restricted to some extent by the presence of the nearby pillar. This restraint to the displacements induced higher confining pressures in the floor, which substantially increased the triaxial strength of the floor when the friction angle was non-zero. Similarly, for the same amount of confinement, the higher the friction angle, the higher the floor triaxial strength. For this reason, when the friction angle of the immediate floor was zero, even though the adjacent pillar was offering similar restraint, the floor strength did not increase a for a non-zero friction angle case.

It has been noted above that the effect of adjacent pillars has more positive effect on the bearing capacity of a

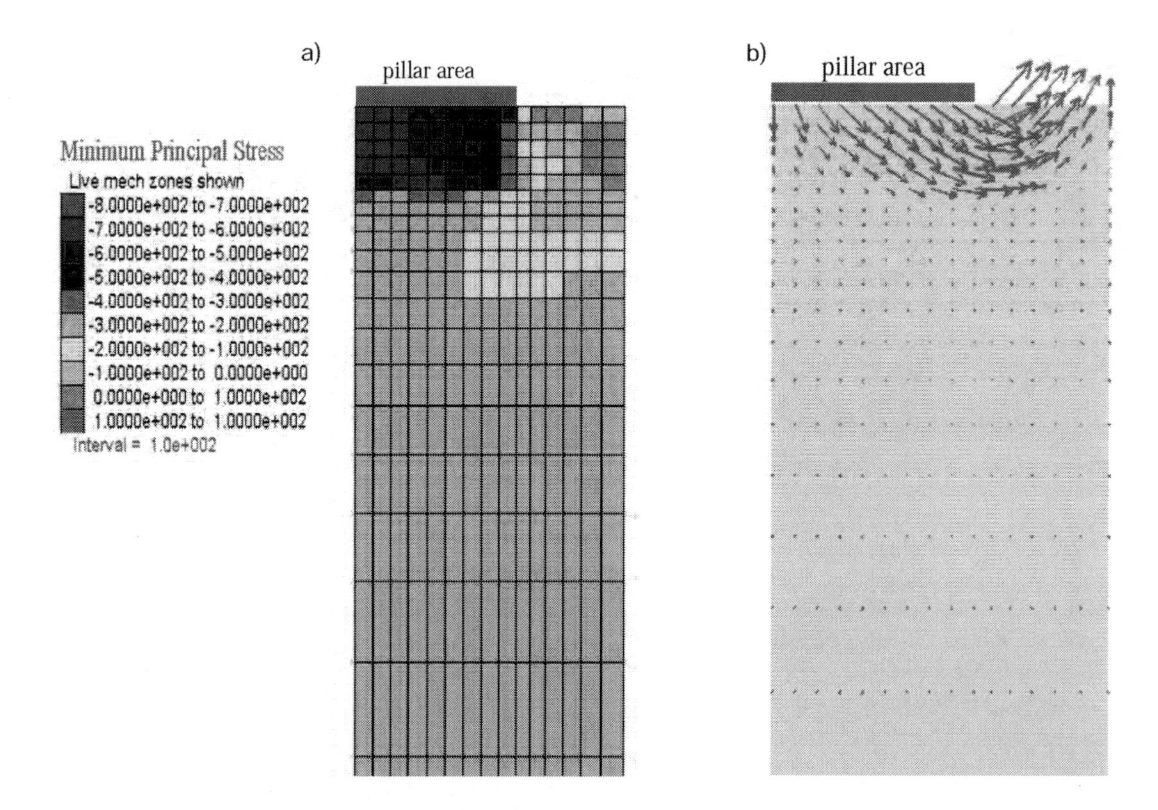

Figure 6.—Model results for multiple strip pillars: (a) contours of minimum principal stress (negative numbers indicate compression); (b) displacement vectors at the limit state.

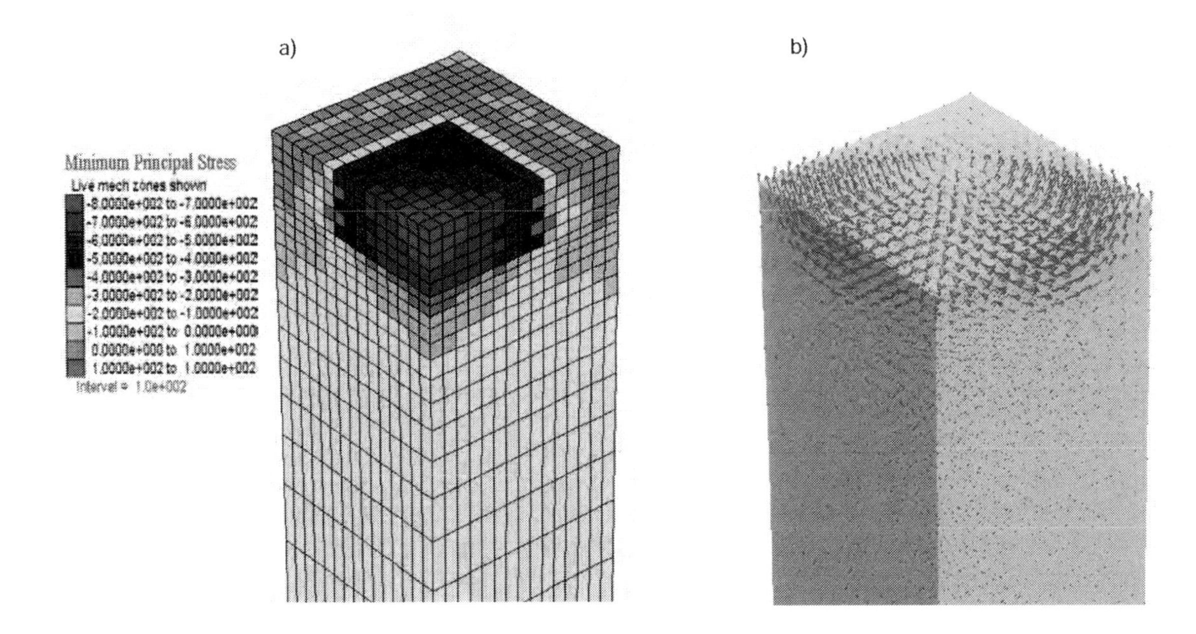

Figure 7.—Model results for multiple square pillars: (a) contours of minimum principal stress (negative numbers indicate compression); (b) displacement vectors at the limit state.

strip pillar than a square pillar. The reason for this difference can be seen by comparing the minimum principal stress and displacement vector plots shown in Figure 6 with those in Figure 7. In both figures, the model conditions were exactly the same except for the geometry of the pillar. The results show that in the case of a square pillar, the amount of floor movement in the entry adjacent to the pillar keeps decreasing from the midportion toward the intersection. As a result, the horizontal confinement generated in the floor keeps decreasing from the middle of the pillar to the intersection area, as seen in Figure 7a. Since the overall confinement of the floor below the pillar is lower for the square pillar, for the same friction angle value, a strip pillar will have higher bearing capacity, as shown by the values in Tables 2 and 3.

CONCLUSIONS

Based on the discussions presented in this paper and the numerical modeling results, the following conclusions can be made:

- When all of the geometric and material property aspects of a typical coal mine bearing capacity problem are considered, it is impossible to provide a closed-form solution for floor strength estimation.
- Limited numerical modeling studies were conducted in the past to study the coal mine floor stability problem. The past studies were limited to explain some site-specific floor behavior or were based on a few parametric studies.
- The nonlinear numerical modeling methodology adopted in this paper reproduced theoretical solutions accurately.
- This paper demonstrates the strength of the numerical modeling approach to estimate floor strength and to shed some light on the possible errors incurred by using conventional bearing capacity theories borrowed from the soil mechanics literature.
- The modeling exercise showed that nonuniform pillar stress distribution may not significantly alter the ultimate bearing capacity value compared to the rigid plate loading. With nonuniform stresses, however, several localized failures occur in the stress concentration zones before the entire floor below a pillar reaches its limit state. Therefore, limited floor heave in underground openings need not necessarily imply floor failure below the whole pillar.
- Interaction between adjacent pillars will positively impact the floor-bearing capacity only if the friction angle of the weak floor material is nonzero. Rectangular pillars will gain more from their neighbors than square pillars of the same width. The dramatic increase in floor strength with friction is related to the increase in the friction-related triaxial strength of the floor material when the adjacent pillar restricts the lateral displacement of the floor.

It is the author's opinion that there is no need to conduct an exhaustive number of parametric studies to develop some "design equations" based on modeling. This is due to the widespread availability of advanced modeling software tools, which make the task of running a few site-specific bearing capacity problems a trivial matter.

REFERENCES

Bandopadhay S [1982]. Analysis of soft floor interaction on underground mining at a western Kentucky mine [Thesis]. Carbondale, IL: Southern Illinois University.

Bhattacharyya AK, Seneviratne SP [1992]. A study of floor heave in the mines of the southern coalfields of New South Wales by two dimensional finite element modelling. In: Proceedings of the 11th International Conference on Ground Control in Mining (Wollongong, New South Wales, Australia), pp. 457–482.

Chandrashekhar K [1990]. Effects of weak floor interaction on underground room-and-pillar coal mining. [Dissertation]. Carbondale, IL: Southern Illinois University.

Chugh YP, Pytel WM [1992]. Design of partial extraction coal mine layouts for weak floor strata conditions. In: Iannacchione AT, Mark C, Repsher RC, Tuchman RJ, Simon CC, eds. Proceedings of the Workshop on Coal Pillar Mechanics and Design. Pittsburgh, PA: U.S. Department of the Interior, Bureau of Mines, IC 9315, pp. 32–49. NTIS No. PB92-205509.

Deb D, Ma J, Chugh YP [2000]. A numerical analysis of the effects of weak floor strata on longwall face ground control. SME preprint 00-134. Littleton, CO: Society for Mining, Metallurgy, and Exploration, Inc.

Rockaway JD, Stephenson RW [1979]. Investigation of the effects of weak floor conditions on the stability of coal pillars. U.S. Bureau of Mines contract No. J0155153. NTIS No. PB 81-181109.

Terzaghi K, Peck RB [1967]. Soil mechanics in engineering practice. 2nd ed. New York: John Wiley and Sons.

Vasundhara LG [1999]. Geomechanical behaviour of soft floor strata in underground coal mines [Dissertation]. Sydney, Australia: University of New South Wales.

Vasundhara LG, Byrnes R, Hebblewhite B, Martin S [2001]. Long term stability of mine workings in soft floor environment: geomechanical investigations at Cooranbong Colliery. In: Peng SS, Mark C, Khair AW, eds. Proceedings of the 20th International Conference on Ground Control in Mining. Morgantown, WV: West Virginia University, pp. 180–183.

Yavuz M, Iphar M, Aksoy M, Once G [2003]. Evaluation of floor heaving in galleries by numerical analysis. In: Proceedings of the Application of Computers and Operations Research in the Minerals Industries. South African Institute of Mining and Metallurgy, pp. 185–190.

IT IS BETTER TO BE APPROXIMATELY RIGHT THAN PRECISELY WRONG: WHY SIMPLE MODELS WORK IN MINING GEOMECHANICS

By Reginald E. Hammah, Ph.D.,[1] and John H. Curran, Ph.D., P.Eng.[2]

ABSTRACT

This paper argues that, due to challenges such as great uncertainty and presence of ill-posed problems, simple models are well suited to mining geomechanics. It builds its case by defining what models are, outlining the usefulness of simple models and explaining how they can be developed. The paper explains that models are necessarily incomplete representations of real-world behavior. The strategy it advocates for constructing a simple model requires a bottom-up approach—starting with the simplest possible model and growing it to capture the essential features of phenomena of interest. The paper calls for engineers to view models for what they really are—tools of the trade, not unlike the physical tools of the sculptor.

INTRODUCTION

In science and engineering, accuracy is the degree to which a measurement or calculated quantity matches its "true" value. Precision is a closely related, but different concept. It is the degree to which repeated measurements or calculations produce same or similar results. It is possible, for example, for a calculation to produce inaccurate but precise answers. This would occur if the answers are consistently close to each other, but are in reality far from being correct.

The first part of the paper's title, "It is better to be approximately right than precisely wrong," is a quote that has been variously attributed to John Maynard Keynes and Warren Buffett. In modeling, it means that although it may be possible to calculate something very precisely, the result may be meaningless if the underlying model, however elaborate, is incorrect. The result may be precisely wrong! In this case, you would be better off with an approximate answer from a simpler model that better represents the real situation.

We will argue in this paper that the quote succinctly describes the case of mining geomechanics. We believe that computer models are fundamentally essential to geomechanics. The paper seeks to emphasize exactly what models are, what they can be used for, and how they can serve our purposes.

The paper will describe three broad challenges in mining geomechanics that make it imperative to prefer simple, approximate models over more complicated, precise ones. These include ill-posed questions and the ubiquitous presence of large uncertainty in mining.

The paper will discuss why simple models are well suited to answering mining geomechanics questions and why complicated models must be avoided at the start of the modeling process. It will outline what constitutes simple models, why they must be used solely as tools, and describes a simple strategy for developing such models.

As part of the effort to justify the use of simple models, the paper will examine lessons we can learn from a very common pest, the cockroach, which has survived for many millennia using seemingly simple models of the environment. Parallels will be drawn between a parable about a superaccurate map and the application of numerical modeling to the problems of mine geomechanics.

One of the key issues emphasized in this paper is the role of models as tools, not unlike the hammer and chisel of the sculptor. Engineers must use modeling the way Michelangelo used sculpting tools to express his vision of the masterpiece *David*.

The paper will illustrate some of the principles advocated through an example in which simple models were used to develop a solution to an ore extraction problem.

WHAT IS A MODEL?

We will begin by describing what a model is. As in science, knowledge and understanding of phenomena in engineering are often embodied in the form of models. As a result, the creation and modification of models is integral to engineering. Engineers use models to predict and control behavior and to develop technologies to satisfy the demands of society.

What then is a model? A model can be defined as a representation of a system that allows us to investigate the behavior and attributes of the system and, sometimes, to predict outcomes of the system under different conditions. The representation is usually—

1. A physical model, such as an architect's model of a building; or
2. An abstraction, such as a set of equations or a computer program.

In this paper, by "model" we mean an abstraction in the form of computer software.

[1]Rocscience, Inc., Toronto, Ontario, Canada.
[2]University of Toronto, Lassonde Institute, Toronto, Ontario, Canada.

Incompleteness of a Model

By necessity, models are incomplete representations of the real world [Mandelbrot and Hudson 2004; Derman 2004]. If a model were to include every aspect of the real world, it would no longer be a model. This is illustrated by the one-paragraph short story entitled ***On Exactitude in Science*** authored by Argentine writer Jorge Borges [1975]. In the story, the cartographers of a fictional empire attained such perfection that they created a map "whose size was that of the Empire, and which coincided point for point with it." Of course, the map was so impractical (it took greater effort to use the map than to actually move around the empire) that succeeding generations abandoned it to the "Sun and Winters."

To create a model, we always make some assumptions about the phenomenon we are representing and the relationships between the different factors that explain the real-world behavior. We strive to include factors that affect behavior and exclude those we deem are not essential. As a result of our assumptions and exclusion of factors, our models are always only approximations, and their results are always estimates. It is good practice, therefore, for us to develop a feel for how far off these estimates are or can be [Poundstone 2005]. We should never take modeling results for granted, but always ask probing questions.

What Can We Accomplish With Models?

Models allow us to attain many useful ends. These include:

1. Development of understanding
2. Proper formulation of questions
3. Reasonable approximation of behavior and provision of meaningful predictions
4. Aid to design of solutions and decision-making

It will be discussed later that ill-posed questions constitute a big challenge in mining geomechanics. A most powerful use of modeling tools is the proper formulation of questions. It has been said that a problem well stated is a problem half-solved [Hubbard 2007]. Models permit us to perform "what if" analysis, which are experiments with different inputs, assumptions, and conditions. Answers to these questions can often lead to the correct diagnosis of problems of key behaviors.

Through the insights they yield, models also help us to reduce uncertainty. Successful modeling does not have to eliminate uncertainty. By merely reducing uncertainty, especially when its costs are much less than the costs of the problem, modeling is often worthwhile. In some cases, models can be explicitly used to assess the likelihood of events and to help formulate plans for coping with such events.

Models Are Tools

A "tool" refers to any device used to perform or facilitate work. Just as a handtool might be used to fix a physical object, computer models can be used to accomplish a task. They are tools of engineering just as hammers and chisels are tools for sculptors. They are tools in the sense that they allow us to explore problems in exhaustive detail without having to do the lengthy and involved calculations. The computer and software do all the drudge work, which enables us to analyze and design.

Through vision and skillful use of the hammer and chisel, Michelangelo sculpted the masterpiece ***David.*** Likewise, our models do not in themselves solve problems. We use them to answer questions.

CHALLENGES OF MINE GEOMECHANICS

Engineering can be defined as the process of providing solutions to the problems of clients as efficiently as possible based on the resources (budget, personnel, time, data, etc.) available to them. The third part of this definition is about challenges. Although these challenges may not be unique to mine geomechanics, they feature strongly in this field. We will examine three of the most important categories of challenges to effective mining geomechanics.

Large Uncertainty

Mining is carried out in the geological environment, which offers one certainty—uncertainty. For our purposes, uncertainty [Hubbard 2007] is defined as the—

- Lack of complete certainty, the fact that the "true" state or outcome is unknown
- Existence of more than one possibility; or
- Chance of being wrong

It gives rise directly to risk—the situation in which some of the possibilities involve loss, catastrophe, or other undesirable outcomes [Hubbard 2007].

In the geological environment, the likelihood of encountering unanticipated conditions is almost always high. The complex behavior of geologic materials and the distribution of their properties in space also do not lend themselves easily to investigation or measurement. Consequently, rock mechanics modeling has been characterized as belonging to the data-limited categories [Starfield and Cundall 1988] of Holling's [1978] classification of modeling problems.

As has been argued by others [Starfield and Cundall 1988], we believe that it is dangerous to apply the methods of exactitude to mining geomechanics problems. Unfortunately, however, the elegance of elaborate models has so fascinated many engineers that answers have been sought that fit models rather than conform to reality in an uncertain world.

Ill-Posed Problems

Geomechanics problems in mining often come to us as questions, ill-posed [Starfield and Cundall 1988] by clients who know they have difficulties, but which they cannot always articulate. Ill-posed problems often have one or more of the following characteristics (adapted from [Denker 2003]):

1. Underspecification or the absence of crucial information that somehow must be determined;
2. Unconnected pieces of information that require understanding in order to determine what is important and what must be ignored;
3. Inconsistent, conflicting, or contradictory information, as a result of which solutions cannot be envisaged, even in principle;
4. Uncertainty as to what solution method or approach must be applied;
5. Ambiguity or the possibility of different answers, depending on what assumptions are used; and/or
6. Intractable answers that exist in principle, but which we have no reasonable ways of determining

The first battle in many mining geomechanics situations, therefore, is to understand what the problem is. After doing this, we are better placed to provide adequate answers. John Tukey, the renowned statistician, once wrote: "Far better an approximate answer to the right question, which is often vague, than an exact answer to the wrong question, which can always be made more precise [Tukey 1962]."

Limited Resources: Personnel, Budget, Time

It is not uncommon for mining geomechanics engineers to work with limited resources—small budgets, tight deadlines, and insufficient personnel. These constraints can be quite severe. Let us take the situation of a rock mechanics engineer in a mine as an example. He/she has several tasks to fulfill each workday in different parts of the operation. This can leave very little time for reflective thinking or strategic problem-solving.

Personnel who are well trained in the use of modeling tools and equally well grounded in practical geomechanics are scarce. In many firms and companies today, the senior engineers who understand well the practical geotechnical issues that need to be resolved are often not that comfortable with numerical modeling tools. Even when they are, they have only limited time to work with them. As a result, they rely on junior engineers for modeling expertise. Often, however, the junior engineers, although comfortable with software, have not acquired sufficient understanding of real-world geotechnical problems and generally require clearly defined questions.

This leads to a situation in which those building the models may not be sufficiently aware of the weaknesses or assumptions in their models, while those who make the decisions may not fully understand what the models are doing or how they work (similar to the situation in financial risk modeling described by Rebonato [2007]). We will argue later that the best way to rectify this situation is to make modeling tools as simple to use as possible. This affords senior engineers time and opportunity to also "play" with models and more fully use their experience for problem-solving and design.

The Opportunities of Constraints

It is easy to view the challenges to mine geomechanics modeling as negatives. However, they can be looked at positively, as opportunities to innovate. Given fewer resources, we are forced to make better decisions [Salmon 2009]. The only question is: What modeling tools are best for addressing mining geomechanics problems under our constraints? We propose to answer that next.

SIMPLE MODELS AS USEFUL TOOLS FOR MINING GEOMECHANICS

We have determined that models are powerful tools for engineers in the quest to determine the best possible solutions to problems under finite (limited) resources. There are costs (time, effort, and money) associated with modeling itself. If we are to be successful, we must keep those costs low.

Let us briefly revisit the parable *On Exactitude in Science* [Borges 1975] for an important lesson on modeling. The cartographers' point-for-point map, although perfectly accurate, was absolutely impractical. The effort and resources required to create and read the map far exceeded any utility it offered. The parable teaches that, in the practical world, simple and easy-to-use tools can be much more useful than very elaborate ones, which although more accurate, may be too costly or impractical to build or use.

In modeling, simplicity is also referred to sometimes as parsimony. The principles of parsimony require that we take great care to develop computational algorithms and models that use the smallest possible number of parameters in order to explain behavior [Mandelbrot and Hudson 2004]. They encourage us to avoid unnecessary complexity and pursue the most straightforward approaches.

Simplicity applies not only to the concepts (essence of phenomena) captured by models, but also to how straightforward it is to use modeling tools. Easy-to-use tools free engineers from drudgery, enabling them to dedicate brainpower to skeptical probing of what can go wrong. We will explore this issue further in a later section.

Reasons for Simplicity

As we have discussed, large uncertainties exist in mining geomechanics. In addition, questions are commonly ill-posed and must be solved under the tight constraints of time, budget, and human resources. Under these conditions, precise answers are not the most useful. Often, understanding of behavior and interactions between various factors must take precedence.

Given these challenges, it is better to omit some details of a problem or imperfectly cover those details (keep models simple) than to try to cover every conceivable aspect, but create an overly complex model. Models that yield "good enough" answers and help us to make decisions suffice [Mandelbrot and Hudson 2004].

Uncertainty is dealt with through parametric and scenario analysis—the assessment of possible ranges of behaviors through variation of input properties and consideration of different conditions—or statistical methods. These approaches all require development of alternative models, accompanied by multiple computations.

Simple models facilitate the use of such techniques. Through the diversity of assumptions and scenarios that we can consider, simple models can help us to develop designs that are robust to unexpected or unusual conditions. The lack of exact numbers should not be equated to knowing nothing [Hubbard 2007]. The information they give reduces uncertainty in our understanding and helps us improve the quality of the decisions we must make. As we have discussed, we can use such models to better define ill-posed problems and test our knowledge and assumptions.

There are many other reasons for adhering to the principle of simplicity in the development of mining geomechanics models. Every parameter or input included in a model introduces a source of uncertainty since we have to assign it a value. Therefore, keeping parameters to a minimum reduces uncertainty in the solution process.

Simple models and modeling tools are also much easier to understand and explain. They make it easier to think through problems. When we start out with simple models and use increments in our understanding to direct further modeling, we are able to identify unnecessary details that have insignificant effects on the model system.

Although it can be argued that simple models are flawed (but we should remember so is every other model no matter how complicated), they should be judged by how much they explain compared to how many input parameters they require. Viewed this way, their strengths over more sophisticated approaches quickly become evident. They are generally easy to use and manage, and much quicker to compute. As a result, they are often of great merit due to the time savings they afford.

What Is a Simple Model?

Our preceding discussions indicate that the greater the number of simplifying assumptions made about the real-world phenomenon we are studying, the simpler the resulting model. We have concluded, therefore, that the ultimate goal of modeling is to create parsimonious models—models that have great range of explanation using the simplest possible concepts and smallest possible number of inputs.

This brings us to a more formal definition of what constitutes a simple model. It is the simplest description of a complex phenomenon that still captures those features we are interested in. It is the model for which any additional gain in explanatory power by including more assumptions or parameters is no longer warranted by the increase in complexity. The art of modeling then reduces to finding the simplest models that do outrageously good jobs at describing complex phenomena [Rebonato 2007]. It aims to say much with little [Mandelbrot and Hudson 2004].

A Simple Strategy for Building Simple Models

The definition of what constitute simple models alludes to a strategy for building them. The process starts with careful reflection on the problem we are trying to solve. This exercise helps us to be clear about our purpose.

We then proceed to build models from the bottom up. We begin with radical simplifications. If investigation shows that the phenomenon of interest cannot appear at this level of simplicity, we add to this model as parsimoniously as possible. The manner in which we enlarge our model is guided by the understanding we gain from study of the influence of each added assumption (concept) or variable. If we determine an addition to be irrelevant to our particular task, we eliminate it.

This process strengthens our fundamental understanding of the phenomenon we are studying. The care and detail we exercise in constructing our simple model compels us to avoid hand-waving (the failure to rigorously address central issues or the glossing over of important details). It forces us to strive to fill gaps in our understanding. When a model is built from the bottom up, it is deemed to have met its goals the moment it passes the test, "Is it fit for its purpose?"

The Trouble With Complexity and Precision

When we start with a model that is too complex, we can quickly reach the point where understanding is replaced by blind faith. A model that starts off complex—i.e., has many inputs, assumptions, and aspects—actually obscures understanding. When too many details are included before the behavior of the model is appreciated, interactions among its components will not be clearly apparent. Such a model becomes little more than a "black box" that mysteriously converts input values to numbers or charts. As a result, its outcomes are not readily interpreted and are difficult to subject to commonsense tests. Rather than clarify, the model confuses.

Obsessive focus on modeling detail often coincides with fascination with precision. Under the considerable uncertainties and other constraints of mining geomechanics, obsession with details we cannot get right and precision we have no hope of attaining hinder our ability to make decisions. Although the ranges of values from simple models may seem less sharp, they offer the advantage of keeping us "honest and humble" about what we are doing, namely, estimating. They help us ward off the "hubris of spurious precision" [Rebonato 2007].

Lessons From Nature

Studies of living organisms can indicate to us optimal strategies for handling large uncertainties and unexpected changes in conditions. The length of time a species has survived is a good measure of how it has adapted under such conditions [Bookstaber and Langsam 1985; Bookstaber 2007].

The common cockroach is an example of an organism that has survived for many millennia because of its strategy for dealing with unanticipated environmental changes [Bookstaber and Langsam 1985; Bookstaber 2007]. Scientists have determined that it uses very simple or "coarse" rules (models) for deciding what actions to take in response to environmental changes. Given a wide set of possible inputs about the environment, the cockroach ignores most of these details and focuses on a select few.

At first glance, it seems that such an approach is not optimal in the least. However, research has shown that although suboptimal for any one environment, "coarse" rules are far more efficient over a wide range of different environments. This is especially true when some of the changes in these environments are unforeseeable. Coarse rules are much more likely to anticipate risks and bring about necessary adjustments.

The cockroach's use of simple models seems to tell us that precision and focus on the known comes at the cost of reduced ability to address the unknown. When we spend less time focusing on detailed investigation, we can spend more time thinking and reacting to unknown conditions.

A NOTE ON EASE OF USE

From the example above, biology seems to indicate that we must run our simple models with different inputs and assumptions in order to cope well with our uncertainties and constraints. This requires that models be developed and changed or manipulated with relative ease. They must be easier and less expensive to manage than the real world.

User-friendly, intuitive software interfaces make this possible. Given the challenges of mine geomechanics modeling, it can be argued that user-friendly interfaces can have far greater impact on the work of engineers than sophisticated underlying model concepts.

The design of a user interface must consider the productivity of users [Curran and Hammah 2006]. It must ensure a short, gently sloping learning curve. Practitioners are keenly aware that people's time costs more than computers and software. The real cost of a modeling tool, therefore, is not so much purchase price as the user effort it demands.

Intuitive, graphical ways of displaying results are also important since they help engineers make sense of model results. Visual representation of data is satisfying to most users because it helps them to make sense of model results in instinctive ways.

EXAMPLE OF GOLDEN GIANT MINE

The Golden Giant Mine is a gold operation in the Hemlo camp in northern Ontario, Canada. The ore body at the mine consists of a main and a lower zone. The main ore body, which is tabular, has a strike length of 500 m, an average thickness of approximately 20 m, and dips at angles between 60° and 70° [McMullan et al. 2004; Curran et al. 2003; Hammah et al. 2001]. The lower zone is 30–80 m below the main zone. The gold-bearing ore is located along the contact of a metasedimentary rock formation with felsic metavolcanic rocks.

Description of the Problem

Near the main shaft was a pillar that was open above and mined out below. It contained 660,000 tonnes of high-grade ore. The original mine plan was to mine all ore at depth, abandon the shaft below a certain level, and extract the shaft pillar as the final mining block. Analyses indicated, however, that a significant portion of this high-grade ore would be lost unless the shaft pillar was mined at the same time as the deep ore.

At the same time, for more than a decade, this pillar had been a source of concern, particularly due to its proximity to the main production shaft. Preliminary modeling had indicated that the pillar and nearby infrastructure were under significant stresses as a result of mining throughout the Hemlo camp [Bawden 1995]. There were also indications that the stress levels were increasing and would adversely impact the shaft stability.

The task, therefore, was to design an extraction sequence for the shaft pillar that would not jeopardize the shaft's integrity. It was evident that the solution would have to reduce the stress concentrations around the shaft.

Constraints and Information From Prior Experience

There were a number of challenges that constrained the numerical modeling tool(s) that could be used on the project. The overall extents of the mine (stopes, infrastructure, and other excavations) were large and laid out in complex, three-dimensional fashion. Due to the tabular ore body, the stopes were flat-shaped. The infrastructure excavations, on the other hand, were more regularly shaped. In addition, data on stress levels and rock mass properties, especially postfailure parameters, were very scant. There was, however, evidence of high-stress damage in parts of the mine. Lastly, a solution to the problem had to be found in short time.

From prior experience with elastic three-dimensional models of the mine, engineers knew that model zones with stresses >98 MPa corresponded well with zones of observed stress damage.

Simple Elastic Modeling and the Determination of Mining Strategy

A displacement discontinuity-based boundary-element program [Vijayakumar et al. 2000], which readily accommodated the different shapes of excavation, was selected as the analysis tool. It could handle the large extents of the mine and the complex three-dimensional layout. It also afforded the ease of model building and computational speeds for developing a solution within the required time. On the other hand, however, it required representation of the three primary rock mass types—ore body, metasedimentary, and metavolcanic rocks—with only one set of elastic properties.

Despite the simplified assumptions of homogeneous material and elastic behavior, the numerical modeling tool showed three-dimensional stress flow patterns and stress concentrations that matched observations at the mine [Hammah et al. 2001]. It helped engineers understand the influence that excavation layout had on stress concentrations within the mine. Each single model run took about 2–3 hours to compute compared with the 20 or more hours it took with a more detailed (multimaterial) boundary-element program.

Numerical studies with the simplified model helped establish that the excavation of a destress slot could reduce existing stresses near the main shaft and control stresses induced during mining of the shaft pillar. The slot pushed high stresses away from the main shaft into non-ore-bearing rock mass zones. The tool allowed engineers to experiment with several alternative slot geometries (location, dimensions, and excavation sequencing) and extraction sequences for the shaft pillar and deeper-lying ore.

Figure 1 shows a three-dimensional view of the final slot geometry and location. The slot was to be 55 m wide by 58 m high and parallel to the main ore body, with a dip of 60° toward the shaft.

Displacements from the numerical modeling indicated that the destress slot would experience closure on the order of 1 m. This meant that the slot thickness had to exceed 1 m [Hammah et al. 2001]. If this condition was not met, the walls would make contact and significantly reduce the efficiency of the slot.

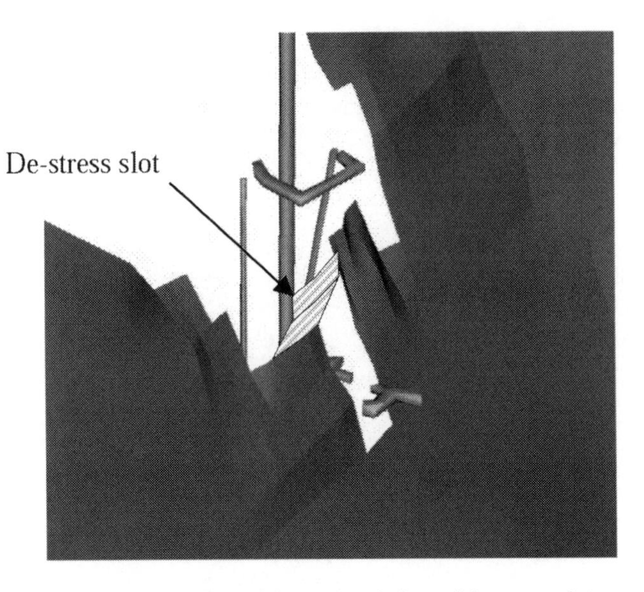

Figure 1.—Three-dimensional view of destress slot geometry and location. The tabular shaded zones represent excavated stopes.

This also led to study of the properties (mainly the Young's modulus or stiffness) of the material for backfilling the slot (there was no way a slot of that size could be left open). For this investigation, a finite-element program, ***Phase***^{2} [Rocscience, Inc. 2009], was used. It could accommodate the multiple material properties integral to the study. Although this program performed only two-dimensional analysis, it was sufficient for this stage of design. The modeling outcomes showed that the material used to backfill the slot had to have very low stiffness—far less than one-fiftieth the stiffness of the host rock.

Real-World Performance of Destress Slot

Excavation of the destress slot, according to the sequence developed from the numerical modeling exercise, began in early 2002 and ended in summer 2003 [McMullan et al. 2004]. The slot was backfilled with a soft paste. Measurement of its performance, which was performed through comprehensive instrumentation, has shown that it met its goals.

CONCLUDING REMARKS

In this paper, we have argued that, in the face of large uncertainties, ill-posed questions, and limited resources, simple, easy-to-use, modeling tools are most practical for mining geomechanics. They facilitate the modeling process. We have also shown that the strategy of building models from the bottom up tends to restrict the creation of complicated, and potentially meaningless, models. We are not advocating simplistic, trivial design and analysis. What we are saying is that mining geomechanics is best served by using the simplest models that fulfill our purposes.

We will end with an analogy from the world of biology—the growth of a sapling [Salmon 2009]. Given enough water and sunshine a sapling will grow. However, with careful pruning—removal of low-hanging branches—during the early stages of development, the sapling will not merely grow but flourish. It will grow faster and become taller and stronger. This is because the pruned sapling will not waste precious resources on growth that does not serve its ultimate purpose.

The same is true of modeling. When we carefully "prune" models and keep them simple, they will help us thrive in solving mine geomechanics questions.

If this paper gives pause for thought any time we must solve mining geomechanics problems, it will have fulfilled its purpose.

ACKNOWLEDGMENTS

The authors would like to thank their colleagues at Rocscience, especially Dr. Thamer Yacoub, for sharing their insights into the topic of modeling.

REFERENCES

Bawden WF [1995]. Hemlo Gold Mines, Inc. Golden Giant Mine class one shaft rehabilitation report. Technical report submitted to Hemlo Gold Mines, Inc.

Bookstaber RM [2007]. A demon of our own design: markets, hedge funds, and the perils of financial innovation. Hoboken, NJ: John Wiley & Sons, Inc.

Bookstaber R, Langsam J [1985]. On the optimality of coarse behavior rules. J Theor Biol *116*:161–193.

Borges JL [1975]. On exactitude in science (Del rigor en la ciencia). In: A universal history of infamy (Historia universal de la infamia). London: Penguin Books. Translated from Spanish.

Curran JH, Hammah RE [2006]. Keynote lecture: Seven lessons of geomechanics software development. In: Yale DP, Holtz SC, Breeds C, Ozbay U, eds. Proceedings of the 41st U.S. Rock Mechanics Symposium (Golden, CO, June 17–21, 2006). Alexandria, VA: American Rock Mechanics Association.

Curran JH, Hammah RE, Yacoub TE [2003]. Can numerical modelling tools assist in mine design? The case of Golden Giant mine. ISRM News J *7*(3).

Denker J [2003]. A rant about story problems including ill-posed problems, and the importance of not always following instructions. Available at: www.av8n.com/physics/ill-posed.htm

Derman E [2004]. My life as a quant. Hoboken, NJ: John Wiley & Sons.

Hammah RE, Curran JH, Yacoub TE, Chew J [2001]. Design of de-stress slot at the Golden Giant mine. In: Proceedings of DC Rocks, 38th U.S. Rock Mechanics Symposium (Washington, DC, July 7–10, 2001).

Holling CS [1978]. Adaptive environmental assessment and management. Chichester, U.K.: Wiley.

Hubbard DW [2007]. How to measure anything. Hoboken, NJ: John Wiley & Sons.

Mandelbrot B, Hudson RL [2004]. The (mis)behavior of markets. New York: Basic Books.

McMullan J, Bawden WF, Mercer R [2004]. Excavation of a shaft destress slot at the Newmont Canada Golden Giant mine. In: Proceedings of the Sixth North American Rock Mechanics Symposium (Houston, TX, June 5–10, 2004).

Poundstone W [2005]. Fortune's formula. 1st ed. New York: Hill and Wang.

Rebonato R [2007]. Plight of the fortune tellers: why we need to manage financial risk differently. Princeton, NJ: Princeton University Press.

Rocscience, Inc. [2009]. Phase^2 v7.0. Finite element analysis for excavations and slopes. [http://www.rocscience.com/products/Phase2.asp]. Date accessed: April 2009.

Salmon F [2009]. Recipe for disaster: the formula that killed Wall Street. Wired Mag, Feb. 23.

Starfield AM, Cundall PA [1988]. Towards a methodology for rock mechanics modelling. Int J Rock Mech Min Sci Geomech *25*(3):99–106.

Tukey JW [1962]. The future of data analysis. Ann Math Stat *33*(1):1–67.

Vijayakumar S, Yacoub TE, Curran JH [2000]. A node-centric indirect boundary element method: three-dimensional discontinuities. Int J Comput Struct *74*(6): 687–703.

AN OVERVIEW OF CALIBRATING AND USING THE LAMODEL PROGRAM FOR COAL MINE DESIGN

By Keith A. Heasley, Ph.D.[1]

ABSTRACT

The LaModel program was developed from the displacement-discontinuity variation of the boundary-element method. It is used mainly to model the stresses and displacements on thin tabular deposits such as coal seams. As with most numerical programs, the accuracy of a LaModel analysis depends entirely on the accuracy of the input parameters. Therefore, the input parameters need to be calibrated with the best available information, either measured, observed, or empirically or numerically derived. In recent years, a common systematic method of developing accurate input parameters for LaModel has been greatly desired by the mining industry. In response to this, several new algorithms have been developed for calibrating the most critical input parameters in the program. In particular, algorithms for calibrating the rock mass stiffness against the expected abutment load extent, the gob properties against the expected amount of abutment/gob loading, and the coal strength based on the expected/desired pillar behavior have been developed and are presented in this paper. Also, to demonstrate the application of LaModel and the new calibration techniques to coal mine design, a case study of a pillar retreat mining section is presented.

INTRODUCTION

Some 15 years ago, the kernel of a laminated overburden model was programmed into a relatively simple program, LaModel, for calculating the displacements and stresses on thin, tabular deposits such as coal mines [Heasley 1998]. The mathematical basis of the program was the displacement-discontinuity variation of the boundary-element method. This approach allowed the program to easily model relatively large areas of the mines (compared to volume-element methods) and to easily model multiple seams. The original desire for the laminated overburden model was that it might provide more realistic displacement calculations than the homogeneous elastic overburden models that were then state of the art. It was even hoped that the laminated overburden model would allow accurate calculation of displacements and stresses at both the seam level and the surface [Heasley and Salamon 1996]. Indeed, the LaModel program did provide more realistic displacement calculations at the seam level than a homogeneous elastic model, and it did provide realistic surface subsidence calculations, although simultaneously calculating accurate displacements and stresses underground and accurate surface subsidence with the same material properties is a goal that the program did not achieve [Heasley 1998; Heasley and Barton 1999].

The original laminated overburden program, LaModel 1.0, had just the basic components for performing the seam displacement and stress calculations. The input overburden parameters, seam parameters, and mine grid were manually typed into a text file with rigid formatting. Then, the DOS-based LaModel program performed the calculations and put the results into a formatted output file for manual analysis. The output consisted of calculations for displacement, stress, multiple-seam stress, overburden stress, and surface-effect stress. This original program provided a 250 × 250 grid for defining the seam geometry and allowed six different material models (linear-elastic coal, strain-softening coal, elastic-plastic coal, linear-elastic gob, bilinear hardening gob, and exponentially hardening gob) for use in 26 different materials for the in-seam element behavior. The original LaModel could analyze up to 4 different seams using a maximum of 20 steps, and it allowed an off-seam plane for calculating remote displacements such as surface subsidence. Also, the original program allowed the input of a topography file for modeling the stress effects of a variable topography [Heasley 1998].

Over the years, LaModel has been upgraded and modernized as operating systems, programming languages, and user needs have evolved. The present LaModel 2.1 program is written in Microsoft Visual C++ and runs in the MS Windows operating system. It uses a forms-based preprocessor for inputting the overburden and seam parameters and a graphical spreadsheet-type interface for creating the mine grid. The preprocessor, LamPre 2.1, also includes a wizard for automatically calculating coal pillars with a Mark-Bieniawski pillar strength and another wizard to assist with the development of "reasonable" gob properties [Heasley and Agioutantis 2001]. Presently, the program can analyze a 1,000 × 1,000 seam grid, and an interface with AutoCAD has been developed to allow the seam and overburden grids to be automatically imported from the mine plans and overburden contours in AutoCAD drawings [Heasley et al. 2003]. The output from LaModel is analyzed with a graphical postprocessor, LamPlt 2.1,

[1]West Virginia University, Department of Mining Engineering, Morgantown, WV.

that quickly and easily generates colored-square plots, cross-sectional graphs, history graphs (which show element changes over the steps), and three-dimensional fishnet plots. For enhanced analysis and display, an interface has been developed to download the LaModel output into AutoCAD and overlay it on the mine map [Wang and Heasley 2005]. In recent years, element and pillar safety factor calculations, and intraseam subsidence and the associated slope and strain calculations have been added to the program [Hardy and Heasley 2006]. Also in recent years, the laminated overburden model has been programmed into a simple two-dimensional version, LaM2D [Heasley and Akinkugbe 2005], and the LaModel output has been combined with geostructural features and geology to create a comprehensive stability mapping system [Heasley et al. 2007; Wang and Heasley 2005].

As a result of the Crandall Canyon Mine collapse in 2007 and the desire in the industry to develop a standardized or best-practice calibration process for LaModel, a number of new enhancements to the program are currently being developed. These include a rock mass stiffness wizard for developing a calibrated lamination thickness, a gob stiffness wizard for developing a calibrated gob modulus, and a strain-softening coal wizard for developing calibrated postfailure coal parameters. An energy release rate calculation, a local mine stiffness calculation, and an overburden fault model are also being added to the program. In addition, with the next release (LaModel 3.0), the grid size will be increased to 2,000 × 2,000, and a number of general enhancements will be made to the grid editor and postprocessor.

LAMODEL CALIBRATION

The devastating pillar failure at the Crandall Canyon Mine in 2007 [Gates et al. 2008] brought coal mine pillar design back to the forefront of mine safety research and raised many questions about the proper techniques for designing mines with numerical models and for developing, or calibrating, input parameters for the models. With LaModel, as with most numerical models, the accuracy of an analysis depends entirely on the accuracy of the input parameters. Therefore, the input parameters need to be calibrated with the best available information, either measured, observed, or empirically or numerically derived [Heasley 2008].

When actually generating a LaModel input file, the geometry of the mining in the seam (or seams) and the topography are fairly well known and can be accurately discretized into LaModel grids with the AutoCAD utility. Therefore, the seam and topography grids do not generally pose an accuracy problem. The most critical input parameters with regard to accurately calculating stresses and loads, and therefore pillar stability and safety factors, are the input material properties, specifically:

- Rock mass stiffness
- Gob stiffness
- Coal strength

These three parameters are always fundamentally important to accurate modeling with LaModel, particularly in simulations analyzing abutment stress transfer (from gob areas) and pillar stability. During model calibration, it is critical to note that these parameters are strongly interrelated, and because of the model geomechanics, the parameters need to be calibrated in the order shown above, since changing the value of one parameter affects the value of the others. The model calibration process as it relates to each of these parameters is discussed in more detail below.

Rock Mass Stiffness

The stiffness of the rock mass in LaModel is determined mainly by two parameters: the rock mass modulus and the rock mass lamination thickness. Increasing the modulus or increasing the lamination thickness will increase the stiffness of the rock mass. With a stiffer rock mass, (1) the extent of the abutment stresses will increase, (2) the convergence over the gob areas and the gob stress will decrease, and (3) the multiple-seam stress concentrations will be smoothed over a larger area. When calibrating for realistic stress output, it is recommended that the rock mass stiffness be calibrated to produce a realistic extent of abutment zone at the edge of the critical gob areas. Since changes in either the modulus or the lamination thickness generally cause a similar response in the model, it is most efficient and logical to keep one parameter constant and only adjust the other. Generally, when calibrating the rock mass stiffness, it has been found to be most efficient to initially select a rock mass modulus (often a thickness-weighted average) and then solely adjust the lamination thickness for the model calibration.

To determine a realistic extent for the abutment zone for calibrating the lamination thickness, it would be best to use specific field measurements of the abutment zone from the mine. However, these field measurements are often not available. In that case, visual observations of the extent of the abutment zone can often be used. Most operations personnel in a mine have a fairly good idea of how far the stress effects can be seen from an adjacent gob. Without any good field measurements or observations, general historical field measurements can be used. For instance, the field measurements used in developing the ALPS and ARMPS pillar design programs indicate that the average extent of the abutment zone (D) at depth (H) (with both terms expressed in feet) for the case study mines was [Peng 2006]:

$$D = 9.3\sqrt{H} \tag{1}$$

or that 90% of the abutment load ($D_{.9}$) should be within [Mark and Chase 1997]:

$$D_{.9} = 5\sqrt{H} \tag{2}$$

Once the extent of the abutment zone (D) at a given site is determined from the best available information, the following equation, derived from the fundamental laminated overburden model, can be used to determine the lamination thickness (t) required to match that abutment extent [Heasley 2008]:

$$t = \frac{2\,E_s\sqrt{12\left(1-\upsilon^2\right)}}{E\,h}\left(\frac{D_n - d}{\ln(1-n)}\right)^2 \tag{3}$$

where E = elastic modulus of the overburden;
υ = Poisson's ratio of the overburden;
E_s = elastic modulus of the seam;
h = seam thickness;
d = extent of the coal yielding at the abutment edge;
and n = percentage of the abutment load.

In Equation 3, the major parameter that is not necessarily known ahead of time is "d", the extent of the yield zone at the abutment edge. This value can be developed by field measurements or observations, or by running LaModel and determining the calculated yield zone for the given conditions. To help get a first approximation of the yield zone extent, one can find the distance into the abutment where the stress gradient implied by the Bieniawski coal strength formula [Mark 1999] is equal to the abutment stress level implied by the laminated overburden [Heasley 2008]. The initial estimate of the yield zone distance (x) is then the solution to the following equation:

$$q\frac{P}{2}\sqrt{\frac{2\,E_s}{E\,\lambda\,h}}\,e^{-\sqrt{\frac{2\,E_s}{E\,\lambda\,h}}\,x} - 2.16\frac{S_i}{h}x - 0.64\,S_i = 0 \tag{4}$$

where q = in situ stress;
P = width of the panel;
S_i = in situ coal strength (psi);
x = distance into the abutment;

and:

$$\lambda = \frac{t}{\sqrt{12\left(1-\upsilon^2\right)}} \tag{5}$$

Gob Stiffness

In a LaModel analysis with gob areas, an accurate input stiffness for the gob (in relation to the stiffness of the rock mass) is critical to accurately calculating pillar stresses and safety factors. The relative stiffness of the gob determines how much overburden weight is carried by the gob and therefore not transferred to the surrounding pillars as an abutment stress. This means that a stiffer gob carries more load and the surrounding pillars carry less, while a softer gob carries less load and the surrounding pillars carry more.

In LaModel, three material models are available to simulate gob behavior: (1) linear-elastic, (2) bilinear hardening, and (3) exponentially strain-hardening. Based on laboratory tests that show that gob is generally exponentially strain-hardening, this is, as a rule, the preferred model for accurately simulating gob behavior [Pappas and Mark 1993]. In the strain-hardening model, the stiffness of the gob is primarily determined by adjusting the "final modulus" parameter [Heasley 1998; Zipf 1992]. A higher final modulus gives a stiffer gob, and a lower modulus value produces a softer gob material. Given that the behavior of the gob is so critical in determining the pillar stresses and safety factors, it is unfortunate that our knowledge of in situ gob properties and stresses is very poor.

For calibrating LaModel, it is imperative that the gob stiffness be calibrated with the best available information. The first thought might be to use laboratory or field measurements of gob material to obtain the final modulus value. However, it is generally very difficult to get an accurate in situ final modulus value directly from small-scale laboratory tests, and direct in situ measurements are quite rare. In lieu of direct measurement, it is recommended that the final gob modulus be calibrated from gob load or abutment load measurements performed at that particular mine. However, these types of field measurements are also fairly rare (and sometimes of questionable accuracy). Also, visual observations are not very useful for estimating abutment loads or gob loads. Therefore, general empirical formulas are quite often the only available information on gob loading.

For coal mining, the most common technique for estimating abutment/gob loads is probably the abutment angle concept (see Figure 1). Using this concept, the abutment load is the weight of the rock contained within the wedge of overburden defined between a vertical line at the edge of the panel and the angled line defined by the "abutment angle" on the inside of the panel [Mark 1990, 1992].

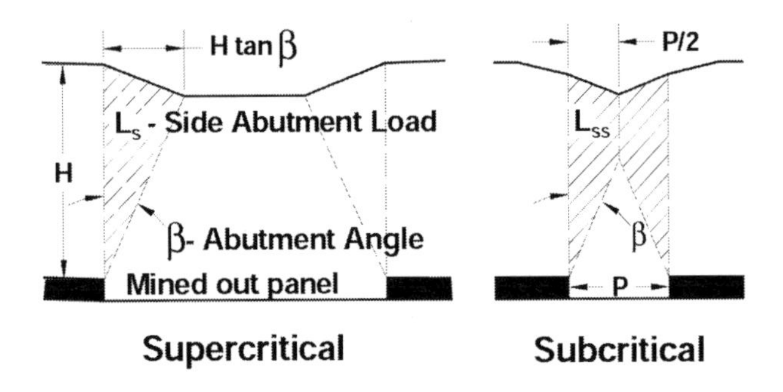

Figure 1.—Conceptualization of the abutment angle.

Using the abutment angle concept (Figure 1), the average gob stress ($\sigma_{gob\text{-}sup\text{-}av}$) for a supercritical panel can be calculated as:

$$\sigma_{gob\text{-}sup\text{-}av} = \left(\frac{H \times \delta}{144}\right)\left(\frac{P - (H \times \tan\beta)}{P}\right) \quad (6)$$

and the average gob stress ($\sigma_{gob\text{-}sub\text{-}av}$) for a subcritical panel can be calculated as:

$$\sigma_{gob\text{-}sub\text{-}av} = \frac{P}{4}\left(\frac{1}{\tan\beta}\right)\left(\frac{\delta}{144}\right) \quad (7)$$

where P = panel width (ft);
β = abutment angle;
H = seam depth (ft);
and δ = overburden density (lb/ft^3).

In the ALPS and ARMPS programs, an average abutment angle of 21° was determined from a large empirical database and is recommended as a default [Mark and Chase 1997; Mark 1990].

When calibrating LaModel, determining a realistic gob or abutment loading is rather difficult without precise field measurements. In general, it has been found that Equations 6 and 7 with an abutment angle of 21° provide fairly conservative estimates of abutment and gob loading and can generally be used [Chase et al. 2002; Heasley 2000]. However, regardless of what value is chosen for the gob/abutment loading, it is clear that a realistic gob loading is critical to realistic model results. This means that the gob stiffness, gob loading, and abutment loading results from a LaModel analysis should be carefully analyzed and seriously debated in any analysis of possible errors. Certainly, gaining a better understanding of the true gob/abutment loading in retreat coal mining is an area for future research.

Coal Strength

An accurate in situ coal strength is another value that is very difficult to obtain, yet remains critical to designing a coal mine and to determining accurate pillar safety factors. The first thought might be to use laboratory measurements of the strength of coal samples to obtain an in situ strength. However, it is very difficult to get a representative sample from the coal seam to test in the lab, and scaling the laboratory values to accurate in situ coal pillar values is not very straightforward or precise [Mark and Barton 1997].

For the in situ coal strength in LaModel, it is strongly recommended to use 900 psi (S_i) in conjunction with the Mark-Bieniawski pillar strength formula as implemented in the coal wizard [Mark 1999]:

$$S_p = S_i\left[0.64 + 0.54\left(\frac{w}{h}\right) - 0.18\left(\frac{w^2}{lh}\right)\right] \quad (8)$$

where S_p = pillar strength (psi);
S_i = in situ coal strength (psi);
w = pillar width;
l = pillar length;
and h = pillar height.

The 900-psi in situ coal strength that is recommended for LaModel comes from the databases used to create the ALPS and ARMPS programs and is supported by considerable empirical data. In the author's opinion, in situ coal strengths calculated from laboratory tests are not any more valid than the default 900 psi due to the inaccuracies inherent to the testing and scaling process for coal strength. If the LaModel user chooses to deviate very much from the default 900 psi, he/she should have a very strong justification, preferably suitable back analyses, as described below, or accurate field measurements.

To determine an appropriate coal strength for LaModel based on a back analysis, the user should analyze a previous mining situation (similar to the one in question) where the coal was close to, or past, failure. The back analysis is an iterative process in which the input coal strength is increased or decreased to determine the value that provides model results consistent with the actual observed failure [Gates et al. 2008]. This back analysis should, of course, use the previously determined optimum values of the lamination thickness and gob stiffness. If there are no situations available where the coal was close to failure, then the back analysis can at least determine a minimum in situ coal strength with some thought as to how much stronger the coal may actually be.

CASE STUDY ANALYSIS

To illustrate how the LaModel program with the calibration techniques described above would be applied to coal mine design, a recent retreat mining analysis is presented here. In this analysis, the active mine was proposing to retreat the #1 Mains (see Figure 2), which were located in between the longwall gobs of the 1 North and 1 South gate roads. The design of the Crandall Canyon Mine area, which collapsed in August 2007, was somewhat similar to the design of the #1 Mains at this mine, where pillars were planned to be extracted in between two longwall gobs [Gates et al. 2008]. Because of this similarity and because of the management's desire to ensure the safety of the mine's operation, a detailed review of the proposed retreat mining pillar plan in the #1 Mains was performed. This review included a comprehensive analysis of the stability of the barrier pillars to the north and south of the #1 Mains and of the section pillars on retreat. Since the stress in this future mining area was a complex combination of the overburden stress, the abutment stresses from retreat mining, and the multiple-seam stress transfer from the overlying mine, the boundary-element program LaModel was an obvious choice for calculating the stress values and safety factors.

In the #1 Mains area, the overburden ranges from 600 to 1,200 ft, with the deepest overburden lying over the middle of the panel (see Figure 2). Also in this area, the active mine is overlain by a previous mine some 400 ft above. The #1 Mains were originally driven in the 1980s with four entries and pillars on 90- by 100-ft centers. Recently, in preparation for extracting these pillars, the section was rehabilitated and an additional entry was driven into the south barrier using 60- by 100-ft pillars. So, in the #1 Mains before pillar retreat, the barrier pillar to the north was 80 ft wide (rib to rib) and the barrier pillar to the South was 140 ft wide (rib to rib). The 1 South and 1 North gate roads adjacent to these barrier pillars were driven with two pillars on 90- by 100-ft centers.

For the LaModel simulation of the #1 Mains, the seams were discretized with 10-ft square elements (which fit the dimensions of the entries and pillars very well) in a 500 × 300 element grid with the model boundary as shown in Figure 2. The model area covers a panel and a half to both the north and south of the #1 Mains section in order to include a full abutment stress from the adjacent panels. Also, the model area extends a fair distance past the mining boundaries of the active section to both the east and west in order to move the edge of the grid outside of the area of influence of the active mining. Symmetric boundary conditions were implemented on all four sides of the model. The grids for both the active and the overlying mine were automatically generated from the AutoCAD mine maps of these mines with some manual oversight. Similarly, the topographic grid was automatically generated from the AutoCAD topographic lines. The topography was discretized with 50-ft elements on a 140 × 100 element grid that extended 1,000 ft beyond all four sides of the displacement-discontinuity grids. The interburden between the seams was set at 400 ft, and the rock

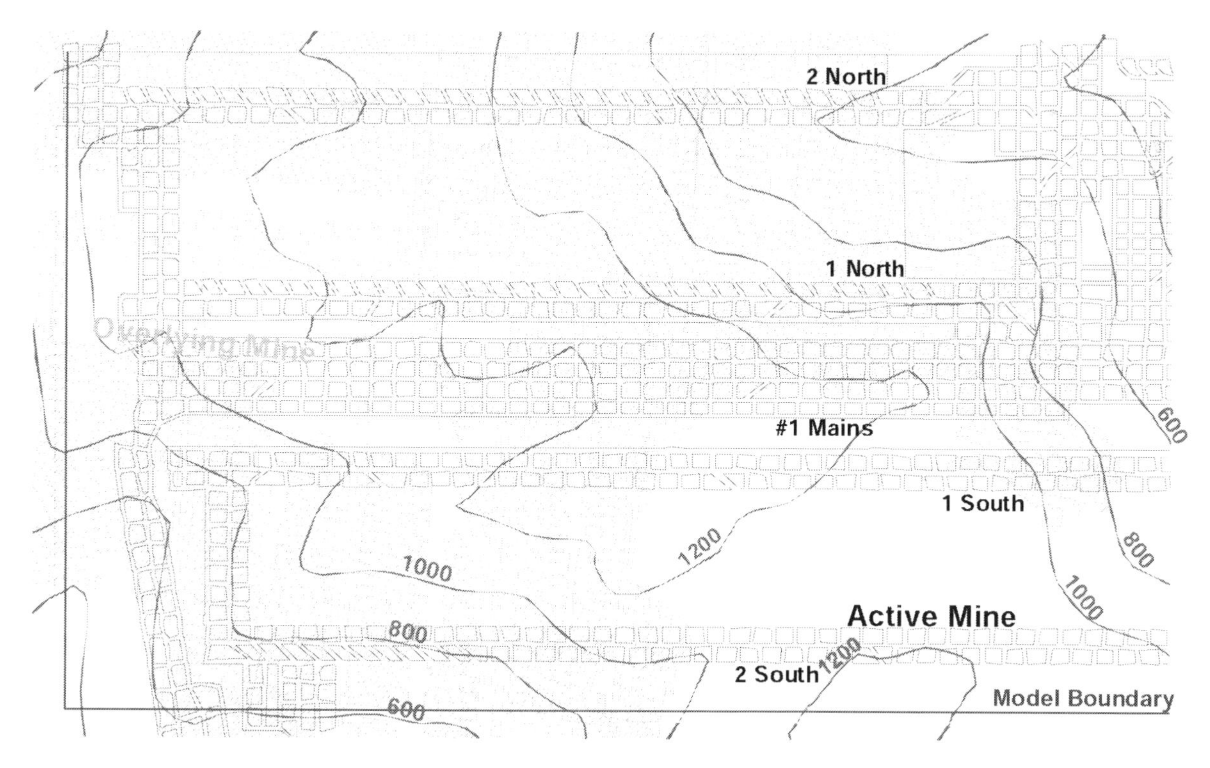

Figure 2.—#1 Mains section at the case study mine.

mass was simulated with an average modulus of 3,000,000 psi. For the active seam, the extraction thickness was set at 6 ft; for the overlying seam, the extraction thickness was set at 4 ft. For both coal seams, the element strengths were determined using an in situ coal strength of 900 psi in conjunction with the Mark-Bieniawski pillar strength formula as implemented in the coal material wizard in LaModel [Heasley and Agioutantis 2001], and the coal elastic modulus was set at 300,000 psi.

In the final LaModel simulation, three mining steps were analyzed: (1) the final development of the #1 Mains, (2) the #1 Mains retreated under maximum cover, and (3) the #1 Mains completely retreated.

Calibration

For calibrating the lamination thickness for this model using Equation 3 above, it can be determined that the lower overburden thickness of 600 ft gives an abutment extent around 120 ft and suggests a lamination thickness around 180 ft and that the highest overburden thickness of 1,200 ft gives an abutment extent around 170 ft and suggests a lamination thickness around 370 ft. Observations at the mine confirm that the 120- to 170-ft abutment extent was fairly close to reality, but in this design analysis, it was also desired to emphasis the multiple-seam stresses on the retreat panel, so a somewhat thinner lamination thickness of 100 ft was ultimately chosen.

For calibrating the gob moduli in this model, the average gob stress values suggested by Equation 7 for 1,200 ft of cover and a 21° abutment angle were examined. For the 700-ft-wide longwall gob, Equation 7 suggests an average gob stress of 500 psi. For the 340-ft-wide #1 Mains gob, the abutment angle concept suggests an average gob stress of 243 psi. Under deeper cover such as in the #1 Mains, it is believed that the abutment angle concept gives lower gob stresses than are actual seen in the field [Heasley 2000], so modeled gob stresses that are a little higher than the calculated values above were deemed reasonable. In order to determine the appropriate gob moduli to use, a single-seam model of the #1 Mains was run with a range of gob moduli from 50,000 to 300,000 psi. From the results of these models, the average gob load versus the gob moduli in both the longwall and #1 Mains was calculated. Ultimately, a final gob modulus of 75,000 psi (an average gob stress of 680 psi) was chosen for the longwall gob, and a final gob modulus of 250,000 psi (an average gob stress of 260 psi) was chosen for the gob in the #1 Mains.

For the coal strength in the model, the 900-psi default value was used. To better understand the appropriateness of this value, a back analysis of a previously mined section of the mine was performed. In this previous section, the cover was similar (1,300 ft) and similar values for the lamination thickness and gob stresses were used. For pillars in the previous section that were adjacent to the gob, a pillar safety factor of 0.76 was calculated, and for pillars one row outby the gob, a safety factor of 1.64 was calculated. According to mine personnel, 2 years after mining, these pillars showed signs of stress and had significant rib sloughing, but were still stable as bleeder pillars and did not require any systematic standing support to maintain the entries. This back analysis result suggests that the 900-psi coal strength was reasonable and may even be a bit conservative. Using the previous section for a relative comparison, the #1 Mains section should be stable if the pillar line safety factors are ≥ 0.76 and 1.64 for the first and second pillar rows from the gob, respectively.

LaModel Output

Once all of the input parameters were developed, as discussed above, and the input file was generated in LamPre, the model was executed in LaModel and it ran for several days on the computer. A number of the important output values from the LaModel run are plotted below in order to analyze the stability of the #1 Mains section.

Overburden Stress

The first output to be analyzed is the overburden stress, as shown in Figure 3. The main reason for plotting and analyzing the overburden stress is to check the overburden input grid for accuracy. As seen in Figure 3, the overburden stress on the active seam is consistent with the overburden contours. The maximum overburden stress on the model is around 1,350 psi at the center of the model, and the overburden stress diminishes as the overburden drops toward the northeast and southwest edges. The "smoothing" of the overburden stress at depth can also be seen in Figure 3. The stress under the valleys is a little higher than the direct overburden value due to the adjacent slopes, and the stress under the ridges is a little lower than the exact overburden value due to the adjacent valleys.

Multiple-Seam Stress

The next output from the model to be analyzed is the multiple-seam stress (see Figure 4). In this plot, only the change in stress on the active seam due to the mining of the upper seam is shown. With the irregular room-and-pillar panels in the overlying seam, the multiple-seam stress exhibits a very irregular pattern. Under the larger pillars between the extracted panels in the overlying seam, the additional abutment stress on the active seam reaches up to 520 psi (an additional 40% over the virgin overburden stress). Under the middle of the extracted panels in the upper seam, the overburden stress on the active seam is partially relieved, up to 760 psi.

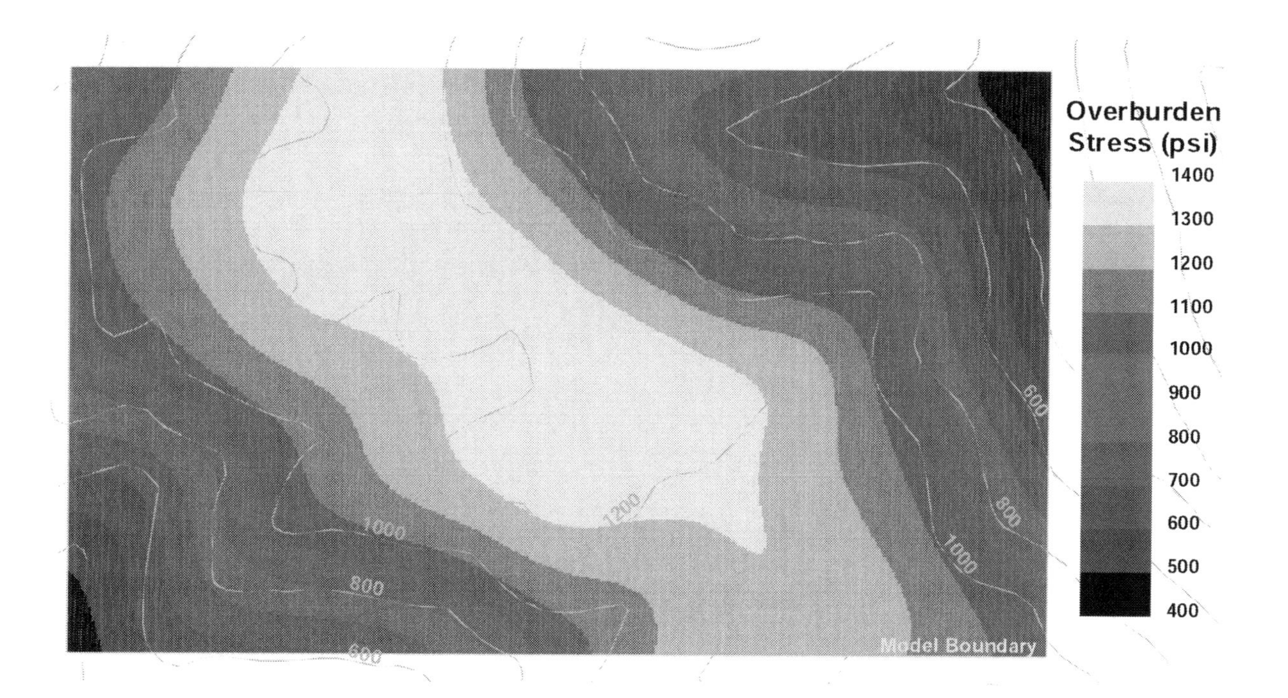

Figure 3.—Overburden stress.

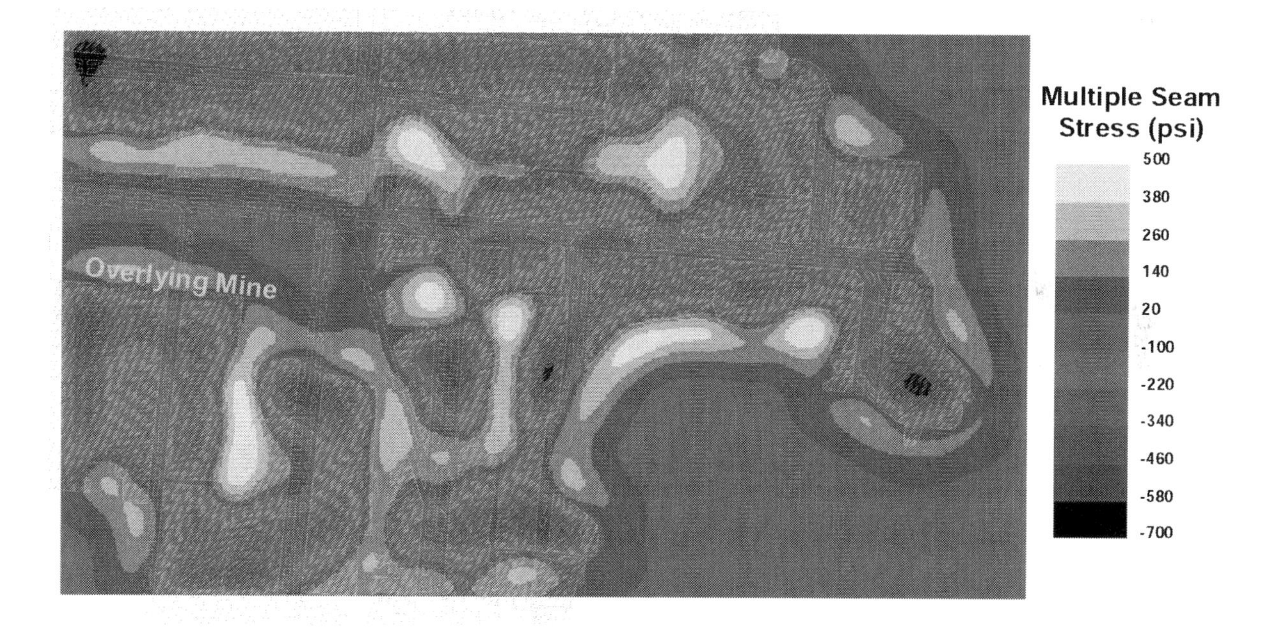

Figure 4.—Multiple-seam stress on the active mine from the overlying mine.

Premining In Situ Stress

The next stress value to be analyzed is the premining in situ stress on the active seam (see Figure 5). This stress value is created by adding the overburden and multiple-seam stresses together, and it shows the effects from both stress components. This premining in situ stress essentially represents the level of stress that will be encountered as the active seam is extracted. As seen in Figure 5, the highest in situ stress is above 1,860 psi (compared to the nominal 1,320 psi overburden stress), and it occurs with a combination of the deepest cover and the high multiple-seam stresses from the abutment pillars in the overlying seam.

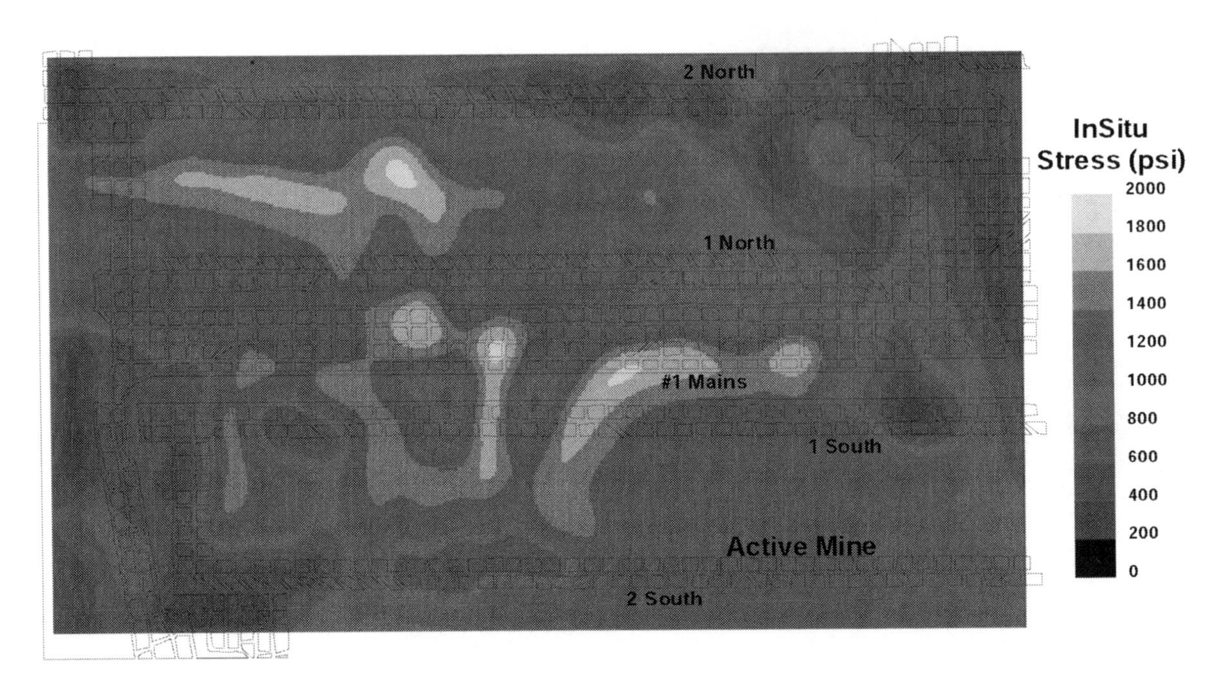

Figure 5.—Element safety factors in the #1 Mains.

Of particular concern with the in situ stress are two very high-stress areas that fall within the #1 Mains—one in the middle of the panel and one on the south side on the outby third of the panel. These high-stress areas will probably not affect the global stability of the section since they are local in size and adjacent to lower-stress areas. However, these local high-stress areas may certainly cause localized high-stress conditions on the retreat pillar line. In the outby third of the panel, it would probably be best from a ground control perspective to pull the pillars from south to north to minimize stresses on the retreat line.

Total Vertical Stress

The next stress item to be analyzed from the LaModel runs is the total vertical stress on the coal seam after the #1 Mains has been retreated about halfway and the retreat line is at the high in situ stress area (see Figure 6). This total vertical stress is essentially the outcome of the overburden, multiseam, and tributary area stresses in the model. It shows the final resultant vertical stress on the in-seam elements. As seen in Figure 6, the highest stresses occur in the core of the gate road pillars, which have considerable abutment stress on them from the adjacent panels. Also, the abutment stress from removing the #1 Mains pillars can be seen on the pillars on the retreat line and on the barrier pillars to the north and south.

To maintain global stability in the #1 Mains, the north barrier pillar between the 1 North gate road and the south barrier pillar between the 1 South gate road need to remain stable. In Figure 6, the abutment stresses from the longwall panels and the #1 Mains retreat section can be seen overriding these barrier pillars. In the south barrier (which is 140 ft wide), the abutment stresses do not increase the stress of the pillar core very much, and this pillar shows considerable stability. However, in the narrower (80-ft-wide) barrier pillar to the north, the abutment stresses are causing 2,000–4,000 psi of stress in the barrier pillar core. Also, the pillars on the retreat line are showing stresses up to 5,000 psi. These high-stress areas cause concern, and the best way to further evaluate the stability in that area is to examine the pillar safety factors in these areas, as presented in Figures 7 and 8.

Pillar Safety Factors

The next outputs to be examined from the LaModel analysis are the pillar and element safety factors. Using the Bieniawski coal pillar formula with a 900-psi in situ coal strength gives a peak strength for each element. This peak strength is then compared with the actual stress on the element to derive an element safety factor. For the pillar safety factors, the LaModel program takes the element safety factor of each element in the pillar and averages them to get the overall pillar safety factor. This algorithm gives a conservative (low) safety factor since a pillar that has not failed completely through the core can have a safety factor less than 1.0 due to the low safety factors of the failed edge elements bringing the average safety factor down.

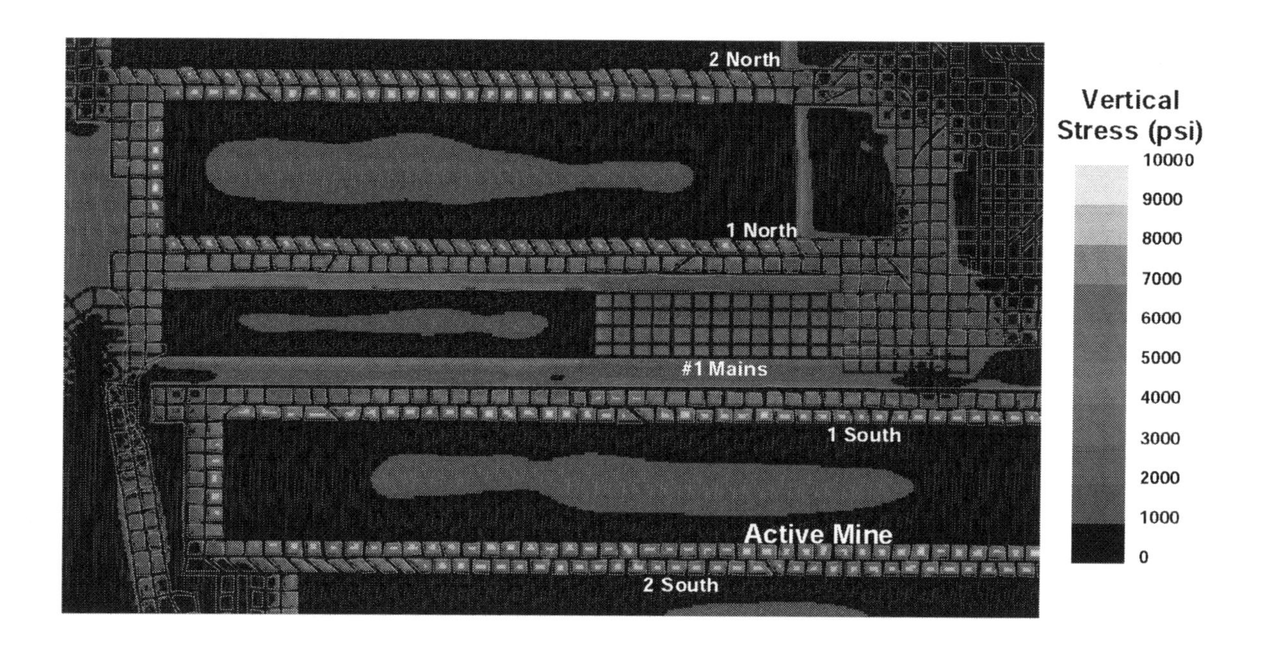

Figure 6.—Vertical stress on the active mine.

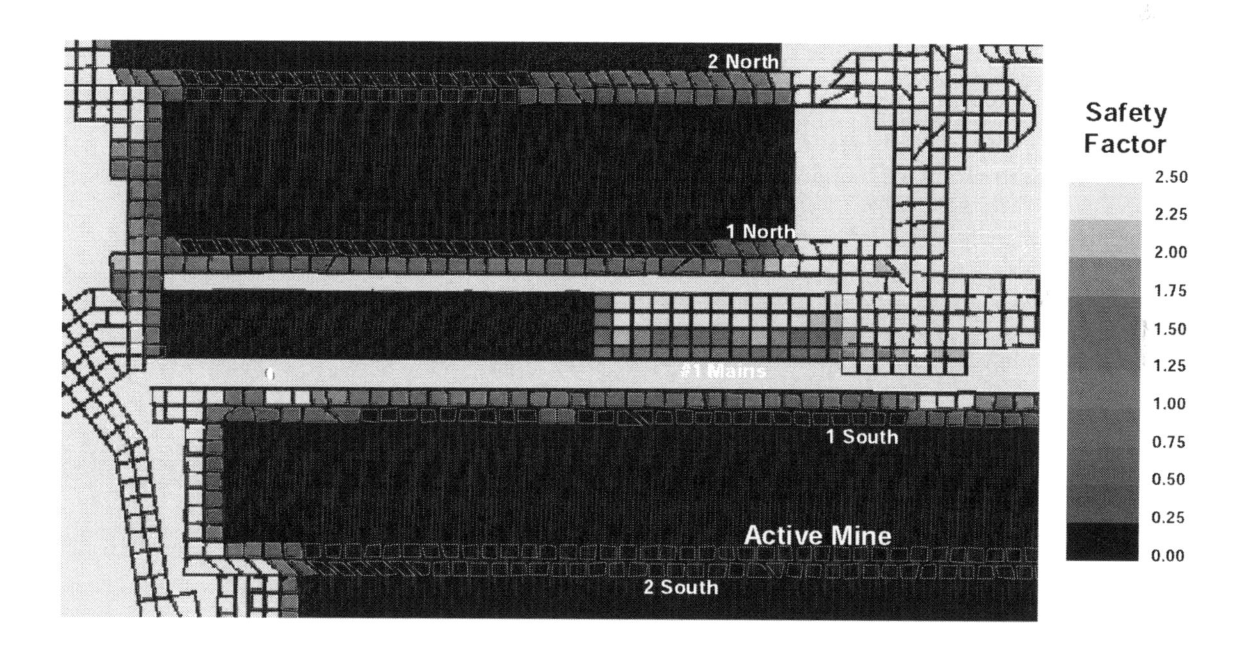

Figure 7.—Pillar safety factors in the #1 Mains.

In the safety factor plots in Figures 7 and 8, many of the gate road pillars (1 North, 2 North, 1 South, and 2 South) show very low safety factors, as would be expected for bleeder and isolated conditions. However, in this analysis, the safety factors of the barrier pillars and retreat line pillars are the keys to global and local stability, respectively.

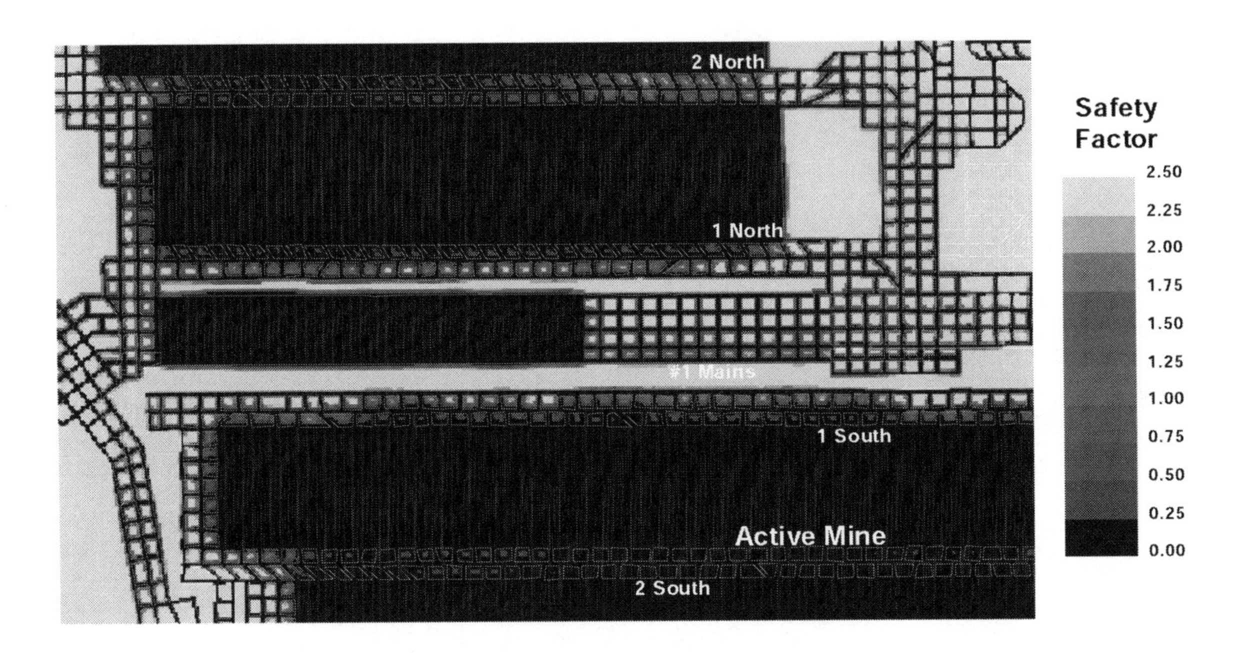

Figure 8.—Element safety factors in the #1 Mains.

As seen in Figure 7, the average safety factor for the north barrier is above 2.25. However, this value is averaged over the entire length of the barrier pillar. In Figure 8, it can be seen that the edges of the north barrier have a considerable number of elements with safety factors less than 1.0, but the core still maintains elements with safety factors greater than 2.0. To determine the practical barrier pillar stability with more detail, the average safety factor for a one-element-wide section of the isolated barrier pillar between the #1 Mains and 1 North was calculated as 1.23. This calculation was performed at the worst location in terms of overburden and multiple-seam stress, and although the safety factor seems low, the pillar should remain stable, in particular, because nearby sections of the barrier have much higher safety factors. Moreover, the analysis is probably a little conservative on the value of the coal strength and gob loading, and failure of the barrier pillar in the gob will not directly affect the stability of the retreat line pillar. The average safety factor of the barrier pillar adjacent to the retreat line has a value of 3.72, which is well above recommended values.

Next, the local stability of the retreat line is analyzed. From Figures 7 and 8, the average safety factor for the middle pillar in the first row of the retreat line in #1 Mains can be calculated as 0.91, and the safety factor of the middle pillar in the second row can be calculated as 1.82. It is not unreasonable to have a safety factor less than 1.0 for the first row of the retreat line. This is a very short-term safety factor; the overall safety factor of the global retreat area (the last couple of pillar rows) is what is critical. In this case, the average safety factor of the last two rows is 1.36. These values compare very favorably with the values determined in back analyzing the previously mined retreat section, which was successful and had a safety factor of 0.76 for the first row of pillars and a safety factor of 1.64 for the second row.

CASE STUDY RESULTS

The main results of this case study, which were presented to the mine management, are the following:

- The worst-case value of 1.26 for the safety factor of the north barrier pillar indicates that the global stability of the #1 Mains section should be adequate.
- The average retreat pillar line stability factor of 1.36 indicates that the local stability of the retreat line should also be adequate.
- In some isolated areas of the #1 Mains, the stresses encountered due to overmining may be about 40% higher than the average in situ stress, and local stability on the retreat line may be a problem.

It was recommended that mine management prepare for localized high-stress conditions in these areas and adjust the mining process as necessary.

Following the LaModel analysis, the #1 Mains section was indeed retreat mined as modeled. There were no problems observed with the global stability of the section and, in general, the pillar retreat line was stable. The mine management planned to leave, and did indeed leave, some pillars under the highest multiple-seam stress areas. The only reported difficulty in the section was some instability at the retreat line after the face had been idled for a couple of weeks.

SUMMARY

The LaModel program, developed some 15 years ago, is based on the displacement-discontinuity variation of the boundary-element method. Because of this formulation, the program is able to analyze the stresses and displacements on large areas of single or multiple tabular deposits such as coal mines. LaModel is unique among boundary-element codes because the overburden material includes laminations, which give the model a very realistic flexibility for stratified sedimentary geologies and multiple-seam mines. Over the years, LaModel has been upgraded and modernized. The current LaModel 2.1 uses a forms-based preprocessor for inputting the overburden and seam parameters, a graphical interface for creating the mine grids, and a graphical postprocessor for generating output plots.

Recent pillar failures in the U.S. mining industry have highlighted the need for a standardized, best-practice calibration process for LaModel. To meet this need, several new algorithms have been developed for calibrating the most critical input parameters in the program. In particular, algorithms have been developed for calibrating the rock mass stiffness based on the expected extent of the abutment load, the gob properties based on the expected gob loading, and the coal strength based on back analysis.

A case study of a pillar retreat mining section was presented to demonstrate the application of LaModel and the new calibration techniques to coal mine design. In this study, the critical input parameters were determined using the recommended algorithms, then the LaModel output was analyzed to determine the ultimate mine stability. During the stability analysis, the utility of using the topographic stress, multiple-seam stress, and safety factor calculations in LaModel was demonstrated and highlighted. Ultimately, the LaModel results were shown to correlate very well with the actual mining results.

REFERENCES

Chase, FE, Mark C, Heasley KA [2002]. Deep cover pillar extraction in the U.S. coalfields. In: Peng SS, Mark C, Khair AW, Heasley KA, eds. Proceedings of the 21st International Conference on Ground Control in Mining. Morgantown, WV, West Virginia University, pp. 68–80.

Gates RA, Gauna M, Morley TA, O'Donnell JR Jr., Smith GE, Watkins TR, Weaver CA, Zelanko JC [2008]. Report of investigation: underground coal mine, fatal underground coal burst accidents, August 6 and 16, 2007, Crandall Canyon mine, Genwal Resources, Inc., Huntington, Emery County, Utah, ID No. 42-01715. Arlington, VA: U.S. Department of Labor, Mine Safety and Health Administration.

Hardy, R, Heasley KA [2006]. Enhancements to the LaModel stress analysis program. SME preprint 06-067. Littleton, CO: Society for Mining, Metallurgy, and Exploration, Inc.

Heasley KA [1998]. Numerical modeling of coal mines with a laminated displacement-discontinuity code [Dissertation]. Golden, CO: Colorado School of Mines, Department of Mining and Earth Systems Engineering.

Heasley KA [2000]. The forgotten denominator: pillar loading. In: Girard J, Liebman M, Breeds C, Doe T, eds. Pacific Rocks 2000 – Proceedings of the Fourth North American Rock Mechanics Symposium (Seattle, WA, July 31-August 3, 2000). Rotterdam, Netherlands: Balkema, pp. 457–464.

Heasley KA [2008]. Some thoughts on calibrating LaModel. In: Peng SS, Tadolini SC, Mark C, Finfinger GL, Heasley KA, Khair AW, Luo Y, eds. Proceedings of the 27th International Conference on Ground Control in Mining. Morgantown, WV: West Virginia University, pp. 7–13.

Heasley KA, Agioutantis Z [2001]. LaModel: a boundary-element program for coal mine design. In: Proceedings of the 10th International Conference on Computer Methods and Advances in Geomechanics (Tucson, AZ, January 7–12, 2001), pp. 1679–1682.

Heasley KA, Akinkugbe O [2005]. A simple program for estimating multiple-seam interactions. Min Eng *57*(4): 61–66.

Heasley KA, Barton TM [1999]. Coal mine subsidence prediction using a boundary-element program. In: Transactions of Society for Mining, Metallurgy, and Exploration, Inc. Vol. 306. Littleton, CO: Society for Mining, Metallurgy, and Exploration, Inc., pp. 99–104.

Heasley KA, Salamon MDG [1996]. A new laminated displacement-discontinuity program: fundamental behavior. In: Ozdemir L, Hanna K, Haramy KY, Peng SS, eds. Proceedings of the 15th International Conference on Ground Control in Mining. Golden, CO: Colorado School of Mines, pp. 111–125.

Heasley KA, Agioutantis Z, Wang Q [2003]. Automatic grid generation allows faster analysis of coal mines. In: Yernberg WR, ed. Transactions of Society for Mining, Metallurgy, and Exploration, Inc. Vol. 314. Littleton, CO: Society for Mining, Metallurgy, and Exploration, Inc., pp. 75–80.

Heasley KA, Petrovich MA, Stone R, Stewart CL [2007]. Mine stability mapping. SME preprint 07-133. Littleton, CO: Society for Mining, Metallurgy, and Exploration, Inc.

Mark C [1990]. Pillar design methods for longwall mining. Pittsburgh, PA: U.S. Department of the Interior, Bureau of Mines, IC 9247.

Mark C [1992]. Analysis of longwall pillar stability (ALPS): an update. In: Iannacchione AT, Mark C, Repsher RC, Tuchman RJ, Simon CC, eds. Proceedings of the Workshop on Coal Pillar Mechanics and Design. Pittsburgh, PA: U.S. Department of the Interior, Bureau of Mines, IC 9315, pp. 238–249.

Mark C [1999]. Empirical methods for coal pillar design. In: Mark C, Heasley KA, Iannacchione AT, Tuchman RJ, eds. Proceedings of the Second International Workshop on Coal Pillar Mechanics and Design. Pittsburgh, PA: U.S. Department of Health and Human Services, Centers for Disease Control and Prevention, National Institute for Occupational Safety and Health, DHHS (NIOSH) Publication No. 99–114, IC 9448, pp. 145–154.

Mark C, Barton TM [1997]. Pillar design and coal strength. In: Mark C, Tuchman RJ, eds. Proceedings: New technology for ground control in retreat mining. Pittsburgh, PA: U.S. Department of Health and Human Services, Centers for Disease Control and Prevention, National Institute for Occupational Safety and Health, IC 9446, pp. 49–59.

Mark C, Chase FE [1997]. Analysis of retreat mining pillar stability (ARMPS). In: Mark C, Tuchman RJ, eds. Proceedings: New technology for ground control in retreat mining. Pittsburgh, PA: U.S. Department of Health and Human Services, Centers for Disease Control and Prevention, National Institute for Occupational Safety and Health, IC 9446, pp. 17–34.

Pappas DM, Mark C [1993]. Behavior of simulated longwall gob material. Pittsburgh, PA: U.S. Department of the Interior, Bureau of Mines, RI 9458.

Peng SS [2006]. Longwall mining. 2nd ed. Morgantown, WV: West Virginia University, Department of Mining Engineering.

Wang Q, Heasley KA [2005]. Stability mapping system. In: Peng SS, Mark C, Tadolini SC, Finfinger GL, Khair AW, Heasley KA, eds. Proceedings of the 24th International Conference on Ground Control in Mining. Morgantown, WV: West Virginia University, pp. 243–249.

Zipf RK Jr. [1992]. MULSIM/NL theoretical and programmer's manual. Pittsburgh, PA: U.S. Department of the Interior, Bureau of Mines, IC 9321.

DEEP COAL LONGWALL PANEL DESIGN FOR STRONG STRATA: THE INFLUENCE OF SOFTWARE CHOICE ON RESULTS

By Mark K. Larson, Ph.D., P.E.,[1] and Jeffrey K. Whyatt, Ph.D., P.E.[1]

ABSTRACT

Software is often used to construct design models used in longwall panel design. These models necessarily reduce the abundant variability found in nature to a simplified representation that, ideally, captures the relevant characteristics of ground response to mining. The choice of stress analysis software is an important step in the modeling process. The importance of this step can be easily overlooked, yet assumptions inherent in modeling software can have a decisive influence. An appreciation of this step is important for both practitioners and users of design model results, particularly when decision-makers are integrating results into design decisions or evaluating the adequacy of design specifications.

This paper examines four different stress analysis programs or tools for analyzing stress around a single longwall panel at 610 m (2,000 ft) depth beneath overburden containing some strong strata. The four tools are: an empirical model that underlies the Analysis of Longwall Pillar Stability (ALPS) program; the three-dimensional displacement-discontinuity program MULSIM/NL; the three-dimensional flexible overburden program LaModel, which has largely superseded MULSIM/NL; and the two-dimensional volume-element program FLAC.

This study examines key aspects of panel simulation, including stress transfer through the gob, stress concentration in abutment ribs, and stress transfer distance into abutments. MULSIM/NL and FLAC results were the most similar. LaModel results varied greatly in terms of peak stress and stress transfer, while ALPS produced the least peak stress. This study also examines the transferability of calibrated input properties between MULSIM and LaModel. For instance, displacement-discontinuity codes have been replaced, at least in part, by LaModel, yet considerable experience exists in calibrated models with older tools. Whether and how this experience can be incorporated into analyses using other tools is a subject of some controversy and must be approached carefully. In all cases, selecting models and properties appropriate to site conditions is extremely important.

[1]Spokane Research Laboratory, National Institute for Occupational Safety and Health, Spokane, WA.

INTRODUCTION

Stress analysis tools such as ALPS, MULSIM/NL, LaModel, and FLAC have been used for some time to evaluate mine layout design. However, each of these tools has underlying assumptions that affect calculated results. This study examines the relative performance of these tools in building a generic design model for a single deep panel in geology typical of western U.S. longwall coal mines. The National Institute for Occupational Safety and Health has been evaluating these tools as part of a research project on control of bump hazards in deep western mines.

The questions posed for this study are as follows:

- Are these tools appropriate for deep western coal mines where the overburden includes one or more strong members?
- What impacts do underlying assumptions have on results?
- Are there ways to translate experience gained with one tool into input for another?

GENERIC MODEL

Larson and Whyatt [2009] used a generic site model to compare the response of ALPS, LaModel, and FLAC to cases with a strong strata member in the overburden (Figure 1). The model's stratigraphic column is typical of deep western coal mines. Elastic properties and strength properties used for each member were within the range of those found in the Wasatch Plateau and Book Cliffs region of Utah [Agapito et al. 1997; Haramy et al. 1988; Jones et al. 1990; Maleki 1995; Maleki 1988; Maleki et al. 1988; Maleki 2006; Pariseau 2007] and are presented in Table 1. These properties were used in the FLAC models. Elastic properties for the displacement-discontinuity codes were averaged from these properties according to the equivalent stiffness method and the weighted thickness method described by Larson and Whyatt [2009]. Only a single panel with 244-m (800-ft) width was considered. The study was expanded to include results of MULSIM/NL for this paper.

In this study, Salamon's [1966] nonlinear reconsolidation model with shale gob parameters as determined by Pappas and Mark [1993] was used as the constitutive law for gob in FLAC. In the case of MULSIM/NL and LaModel, the linearly hardening gob model was used, with parameters fit to the shale gob model used in FLAC.

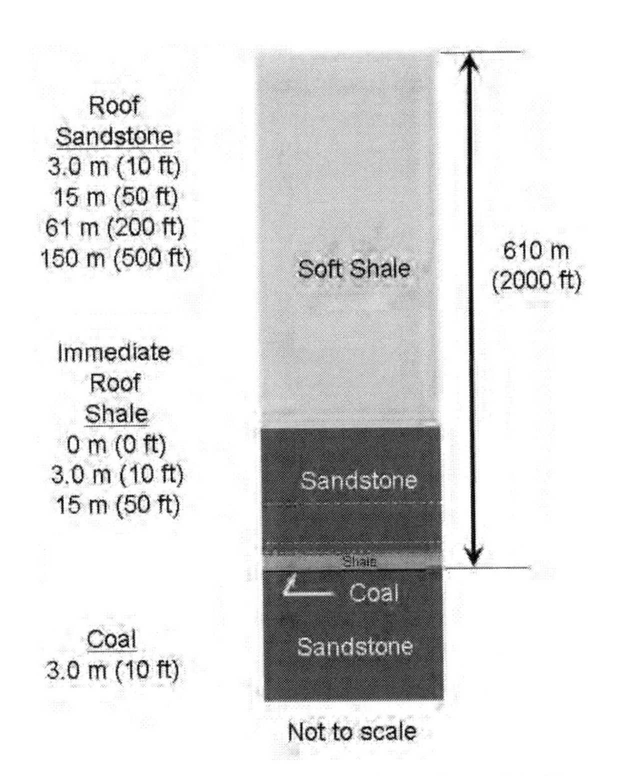

Figure 1.—Stratigraphic column of generic model with indicated thickness of members used in models.

TOOL CAPABILITIES AND ASSUMPTIONS

Numerical stress analysis tools are used in most cases to evaluate mining layout designs. For any given case where there is an excavation and adjacent pillars and abutments, three key points of information are determined by that model. Figure 2 illustrates these three points: (1) the fraction of overpanel weight that is transferred to the abutment versus the gob, (2) the distance into the abutment that the overpanel weight is transferred, and (3) the peak stress in the remaining pillars or abutment and its location.

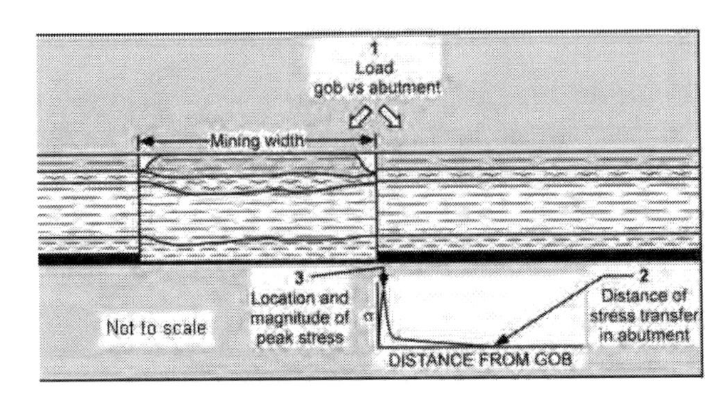

Figure 2.—Vertical cross-section across the width of a panel showing important concepts of stress redistribution resulting from excavation of a single panel.

Each tool simplifies the problem with underlying assumptions. For any specific case, the user must evaluate whether the underlying assumptions of a tool are appropriate. If a software tool is used without considering the underlying assumptions, the tool may give erroneous results. A brief summary of each tool with its capabilities and assumptions follows.

Table 1.—Estimated properties of materials as used in the generic and FLAC models

Property	Soft shale	Sandstone	Shale	Coal
Young's modulus, GPa	10.3	34.6	13.8	3.45
Young's modulus, million psi	1.50	5.00	2.00	0.50
Poisson's ratio	0.35	0.25	0.35	0.30
Density, kg/m^3	2,310	2,310	2,310	1,280
Density, lb/ft^3	144	144	144	80
Cohesion, MPa	20.5	33.8	20.5	7.09
Cohesion, psi	2,970	4,910	2,970	1,030
Friction angle, °	30	25	30	30
Dilation angle, °	5	5	5	5
Tensile strength, MPa	2.07	5.03	6.89	2.07
Tensile strength, psi	300	730	1,000	300
Ubiquitous joint angle, °	0	—	0	—
Ubiquitous joint cohesion, MPa	1.4	—	1.4	—
Ubiquitous joint cohesion, psi	200	—	200	—
Ubiquitous joint friction angle, °	25	—	25	—
Ubiquitous joint dilation angle, °	5	—	5	—
Ubiquitous joint tensile strength, MPa	0.83	—	0.83	—
Ubiquitous joint tensile strength, psi	120	—	120	—

ALPS

Mark [1987] used case studies to empirically calibrate a simple estimate of stress distribution around a retreating coal panel. This method, called ALPS, considers a longwall panel across its width in vertical cross-section. Analysis of Retreat Mining Pillar Stability (ARMPS) is its equivalent for room-and-pillar retreat mining [Mark and Chase 1997]. ALPS considers no geology, only width of the excavation and overburden height to determine a stress distribution on the coal seam. Load on pillars and their stability factors are then calculated. Mark [1990] defines the stability factor as the load-bearing capacity of the pillar system divided by the design loading. A database of stability factors for various cases and their classification of "satisfactory" or "not satisfactory" allows the user to compare the case at hand with many others. Thus, deficiencies in estimation are "corrected" by using experience to define the critical stability factor.

The original ALPS did not take into account the condition of the roof. Molinda and Mark [1994] developed the Coal Mine Roof Rating (CMRR). ALPS stability information was then modified to incorporate roof conditions. Mark et al. [1994] noted that the line

$$\text{ALPS SF} = 1.76 - 0.014\ \text{CMRR} \tag{1}$$

separated "successful" cases from "not successful" tailgate cases in 82% of the case histories in the database.

Advantages of this tool are:

- It is quick and easy to calculate the stability factors.
- It has a large database for comparison with other cases.

The assumptions of this tool are:

- Caving and load transfer fit a simple model. Figure 3 depicts supercritical and subcritical vertical sections showing a wedge volume with unit thickness of the overpanel weight that is transferred to the abutment. That wedge is defined by the angle, β. No differences in geology are directly considered. The overpanel strata between the triangles are assumed to cave, and their weight is fully supported by the gob.
- Mark [1990] found for six cases in the Eastern United States, β ranged from 10.7° to 25.2° and recommended that β be assumed as 21°. That assumption is constant in the ALPS database.

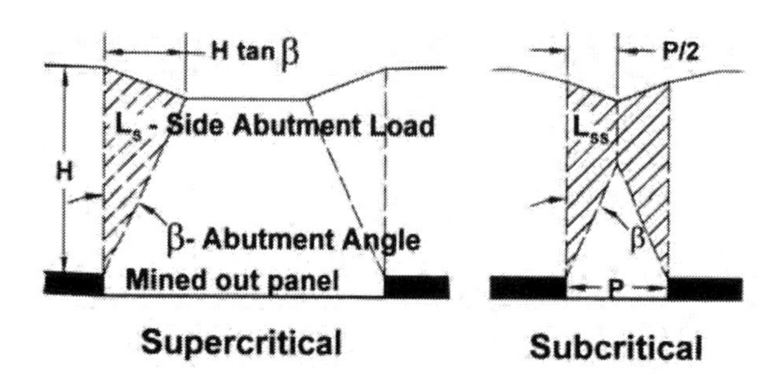

Figure 3.—Vertical cross-section across the width of mined panels showing geometry of supercritical *(left)* and subcritical *(right)* panels (after Heasley [2008a]). The subcritical geometry is used by Mark [1987] in ALPS.

Maximum load transfer distance is represented by the following equation [Pariseau 2007]:

$$D = 9.3\sqrt{H}, \tag{2}$$

where D = maximum load transfer distance (ft);
and H = overburden height (ft).

The vertical stress profile on the seam is represented by

$$\sigma_a = \left(\frac{3L_s}{D^3}\right)(D-x)^2, \tag{3}$$

where σ_a = abutment stress distribution function;
x = distance from the edge of the panel;
and L_s = total side abutment load.

MULSIM/NL

Boundary-element techniques, pioneered in the late 1960s and early 1970s, have a computational advantage over volume-element techniques for problems in infinite or semi-infinite domains in that the system of equations to solve is much smaller for the same problem [Crouch and Starfield 1983]. However, the equations are not sparse, as with volume-element tools, meaning that there are not many zero coefficients in the system of equations that must be solved. The displacement-discontinuity method is a subset of the boundary-element method, which solves the problem of a discontinuity in displacement between opposite surfaces of a crack over a finite length in an infinite elastic medium. Such a solution can be applied to a tabular deposit, such as a coal seam, where the behavior of the deposit is simulated with the crack. The seam is represented by a grid, and each square or block in that grid is assigned its own set of properties, strengths, and constitutive law. Constitutive laws for seam elements include linear elastic, strain softening, elastic-plastic, bilinear hardening, strain hardening, and linear elastic gob. The off-seam material is isotropic elastic only.

Crouch and Fairhurst [1973] developed software implementing the technique for mine structural analysis. St. John [1978] used the technique in his code, EXPAREA. Sinha [1979] used the technique in the development of three computer programs, one of which (MULSIM) analyzes cases of multiple, parallel seams. He included the ability to subdivide coarse blocks into finer mesh so that the scheme was computationally more efficient. Beckett and Madrid [1986, 1988] developed additional features for MULSIM (their version was called MULSIM/BM), such as additional seam materials like gob, pack walls, and cribs; graphical development of grids; and an increase in the number of coarse blocks and the number of blocks that could be subdivided into finer mesh. Donato [1992] converted MULSIM/BM to a PC environment. Itasca Consulting Group added the ability to consider multiple mining steps [Zipf 1992b]. Zipf [1992a,b] added nonlinear seam materials such as strain-softening, elastic-plastic, bilinear hardening, and strain-hardening (MULSIM/NL).

MULSIM/NL can handle up to four parallel seams dipping at some angle to horizontal as specified by the user. Initial three-dimensional stress conditions and stress gradients with depth are specified. Once the system of equations is solved iteratively, the full stress tensor and displacement vector components are output for each element and at user-specified locations in the surrounding ground.

The DOS version (or Zipf version) of the tool uses coarse mesh to streamline calculations. For problems in this study (68 × 68 coarse blocks with fine mesh in the middle 50 × 50 of the coarse blocks), a stress threshold of 1 psi served as the equilibrium convergence criterion so that the number of iterations typically ranged from 250 to 350.

The Windows version (or Heasley version) of the tool lacks the coarse mesh, but allows a 400 × 400 fine mesh. Meshes of this size were initially used in this study, but all iterations were stopped at 90. Use of this version in this study was eventually abandoned because of long run time.

Advantages of this tool are:

- Calculation time with the Zipf version is relatively short.
- The full stress tensor is output.
- Nonlinear in-seam behavior is available, as mentioned earlier.

Assumptions of this tool are:

- Overburden and underburden behavior can be adequately simulated with a one-material, elastic medium.
- Interaction between an elastic off-seam material and a nonlinear seam will adequately and realistically simulate gob-roof displacement and caving behavior. In short, the roof does not fail and cave. Instead, elastic sag of the overburden and appropriate in-seam gob constants can provide realistic load to the gob.

LaModel

Heasley [1998] developed LaModel, a displacement-discontinuity modeling tool that uses the thin-plate or lamination formulation [Salamon 1991]. Each layer is separated by parallel, frictionless joints in the overburden and underburden at even intervals specified by the user. Heasley also included the same nonlinear models used in MULSIM/NL. The frictionless joints make the overburden less stiff, increasing closure of excavated areas of the seam. Heasley found this formulation tracked stress distributions in a coal seam and provided a better match to subsidence observations than an elastic model [Heasley 1998]. However, he cautioned that the user must calibrate for displacement or stress, and calibrating for one entity may not provide realistic results for the other [Heasley 2008b].

Solutions are completed about the same or faster than the Zipf version of MULSIM/NL. Off-seam calculations take a much longer time. Horizontal stresses are not computed. In addition, the seam can only be horizontal. Overburden and underburden are assumed to consist of layers of elastic material interspersed with horizontal, frictionless, cohesionless joints. The result is an increase in mechanical flexibility and the amount of predicted surface subsidence compared to the MULSIM/NL model, which assumes elastic behavior of the overburden.

Heasley [2008b] suggested a method for calibrating a LaModel simulation. He recommended adjusting the overburden Young's modulus so that 90% of the load transferred to the abutment lies within

$$D_{.9} = 5\sqrt{H}, \tag{4}$$

where H = seam depth (ft). Then Heasley recommended using a layer thickness according to

$$t = \frac{2E_s\sqrt{12(1-\nu^2)}}{Eh}\left(\frac{5\sqrt{H}-d}{\ln(0.1)}\right)^2, \tag{5}$$

where E = elastic modulus of the overburden;
ν = Poisson's ratio of the overburden;
E_s = elastic modulus of the seam;
h = seam thickness (ft);
d = extent of the coal yielding at the gob edge (ft);
and H = seam depth (ft), as in Equation 4.

If E_s, v, h, H, and d are maintained constant, then Equation 5 can be reformed to find the product, tE, that is:

$$tE = \text{constant} \qquad (6)$$

In fact, this product is the actual overburden "property," i.e., calculated results will be the same as long as this product is constant.

Advantages of the tool are:

- Calculation time is relatively short.
- Nonlinear in-seam behavior is available.

Assumptions of this tool are:

- Overburden and underburden behavior can be adequately simulated with a one-material, elastic medium with embedded frictionless, cohesionless joints.
- Interaction between an elastic off-seam material with embedded frictionless, cohesionless joints and a nonlinear seam will adequately and realistically simulate gob-roof displacement and caving behavior. In short, the roof does not fail and cave. Instead, elastic sag of the overburden and appropriate in-seam gob constants can provide realistic load to the gob.

EQUIVALENCY OF MULSIM/NL AND LAMODEL

The transition from an elastic overburden mass in MULSIM/NL to overburden with frictionless, cohesionless joints in LaModel begs the question whether MULSIM/NL and LaModel can produce equivalent or similar results for appropriate inputs. Gates et al. [2008, Appendix V], based on Heasley's [1998] formulation, proposed defining equivalent properties by matching midpanel closure. Solutions for the two methods, assuming cracks rather than coal seams, are:

$$s_h(x) = 4(1-\upsilon^2)\frac{q}{E}\sqrt{(L^2 - x^2)}, \text{ and} \qquad (7)$$

$$s_l(x) = \frac{\sqrt{12(1-\upsilon^2)}}{t}\frac{q}{E}(L^2 - x^2), \qquad (8)$$

where s_h = seam convergence of the homogeneous (MULSIM/NL) case;
s_l = seam convergence of the laminated case;
x = distance from the panel centerline;
v = rock mass Poisson's ratio;
t = layer or lamination thickness;
q = overburden stress;
E = rock Young's modulus;
and L = half-width of longwall panel.

Combining these, Gates et al. [2008, Appendix V] found

$$t = \sqrt{\frac{3}{4}}\frac{E_{homogeneous}}{E_{laminated}}\frac{L}{\sqrt{1-\upsilon^2}}, \qquad (9)$$

or

$$tE_{laminated} = kE_{homogeneous} \qquad (10)$$

Gates et al. [2008, p. V-1] present Equation 9 as a method for finding the "required thickness" for translating a calibrated elastic overburden modulus to LaModel overburden properties. More specifically, they state that Equation 9 "could be used to estimate properties that would equate the laminated strata behavior with the homogeneous rock mass used in other boundary element programs" [Gates et al. 2008, p. 115]. This suggests that Equation 9 may provide equivalence beyond closure at centerline of the panel. Such an equivalence would be a valuable link between tools, even though limited to a particular panel width (2L).

A simple test was devised to explore the extent of this equivalence. An elastic model was constructed for both MULSIM/NL and LaModel with layer thickness calculated according to Equation 9. Young's modulus of coal was set high (207 GPa (30,000,000 psi)) to minimize seam contribution to closure. Figure 4 shows the stress and closure profiles for a case with panel half-width set at 122 m (400 ft), Poisson's ratio set at 0.35, and $E_{homogeneous}$ = $E_{laminated}$ = 10.3 GPa (1.5 million psi); and thus t = k = 113 m (370 ft). At midpanel, LaModel calculated closure to be 9.8% higher than MULSIM/NL. This difference likely is a result of effects from element size and edges [Heasley 1998, 2009; Zipf 1992b]. However, the stress profile within 46 m (150 ft) of the panel differed significantly. Thus, this "equivalence" is extremely limited and should *not* be used if abutment stresses are a concern, which is usually the case.

FLAC OR VOLUME ELEMENT

The concept of volume-element discretization for stress analysis has been around for a long time, but has become increasingly important since the advent of computers. For example, the finite-element method was developed in the aerospace industry in the 1950s [Segerlind 1976]. Turner et al. [1956] are credited with being the first to use the method in solid mechanics in 1956. Finite-difference concepts have also been long known to the mathematical world [Dahlquist and Björck 1974]. Dr. Peter Cundall first used the technique in his FLAC computer code to solve solid mechanics problems specifically for geomaterials in 1986.

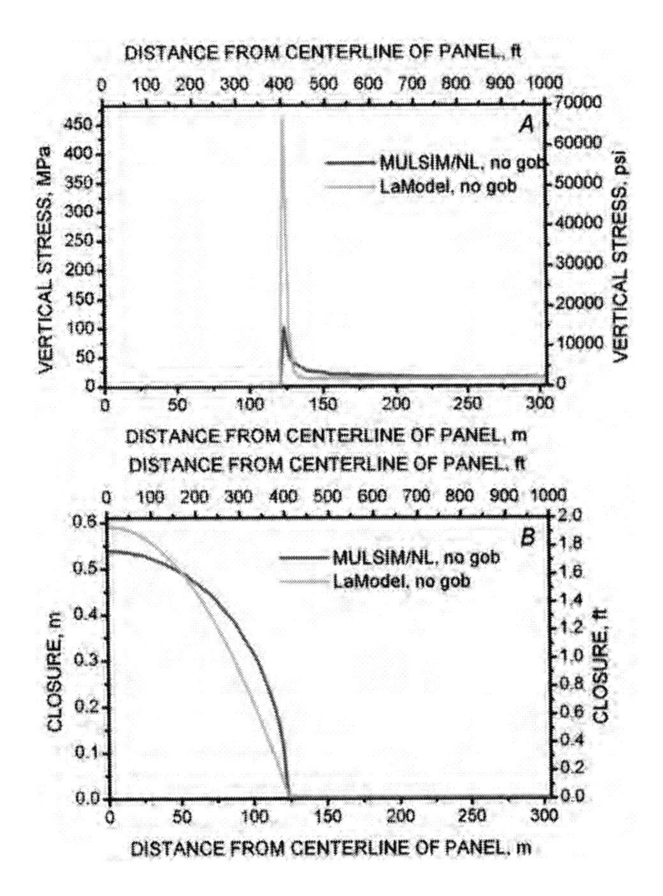

Figure 4.—*A*, Vertical stress profile; *B*, closure profile for a purely elastic case with large Young's modulus for coal. The closures at centerline of panel were expected to be equivalent.

FLAC, as with most volume-element codes, calculates stress for each discretized volume element and displacement at each grid point that defines the volume elements. Each element has its own constitutive law and properties, including the possibility of material yielding and plastic flow. With FLAC version 5.0, used in this study, care must be taken in choosing mesh size and strength properties. If an element exhibits too much localized plastic flow, numerical instability may result. To detect such an impending situation, FLAC stops calculations if any triangular subzone of an element has an area below a threshold fraction of the entire element volume. However, FLAC version 6.0 includes the capability of dynamic meshing, thus eliminating this instability under large plastic deformations.

Volume-element models discretize all modeled space, so boundaries are needed to limit model size. As a result, model boundaries must be far away from the area of interest to avoid influencing results, but close enough to limit problem size. Generally, this method needs a large number of elements and, thus, the size of the model can be very large.

Advantages of FLAC are:

- Each member of the stratigraphic column can be represented according to an appropriate constitutive law, specific elastic properties, and specific strength properties.
- Failure of elements is determined by the code, not by the user.
- The code has an embedded simple computer language, FISH, that permits the user much versatility in model construction, model running, and inclusion of user-defined constitutive laws.
- Complex boundary conditions and initial conditions can be input to the model.

Assumptions of this tool are:

- In the case of a two-dimensional model, the vertical cross-section of the model is far enough from panel ends that a plain-strain condition exists.
- The first-order volume-element response (in the elastic formulation, terms with exponents greater than 2 are neglected) adequately represents material behavior. If beamlike behavior is important in an analysis, then the grid must be fine to calculate accurate deformations and stresses.

GENERIC MODEL RESULTS

The generic model study with each of the four tools considered estimated three aspects of stress redistribution resulting from panel mining, as shown in Figure 2. These are (1) the fraction of overpanel weight that is transferred to the abutment versus the gob, (2) the distance into the abutment that the overpanel weight is transferred, and (3) the peak stress in the remaining pillars or abutment and its location.

Overpanel Weight Transfer to Abutment or to Gob

Figure 5 shows the range of overpanel weight fraction transferred to the abutment for roughly "equivalent" properties. The range for MULSIM/NL was very small—approximately 0.94–0.96 for the whole range of overburden properties used. MULSIM/NL and FLAC results are similar. The larger range of FLAC results was likely caused by various degrees of failure in the set of models. The full range of LaModel results with the same overburden elastic properties (E, ν) was less than that of MULSIM/NL. Stress transfer with LaModel using a lamination thickness on the order of the overburden thickness was clearly not the same as that calculated by MULSIM/NL. The fraction of overpanel weight transferred to the abutment by ALPS for β = 21° was approximately 0.73—clearly below that of FLAC and MULSIM/NL. The upper value of 38° was chosen to test the impact of increased bridging of strata on results.

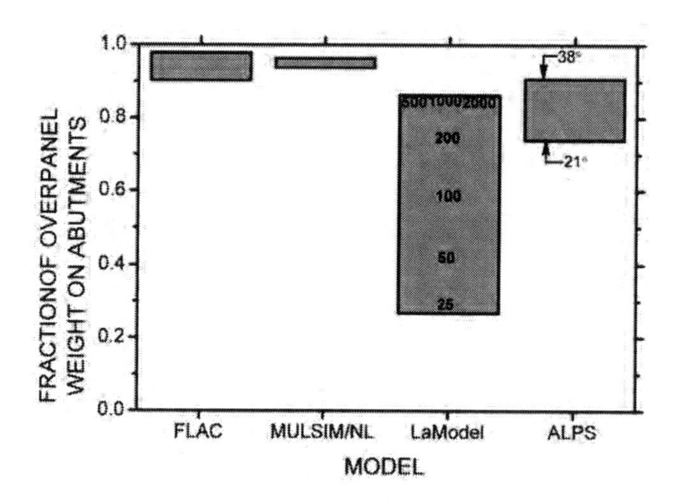

Figure 5.—Chart showing the range of proportion of overpanel weight shifted to abutments for three numerical modeling tools and ALPS for a panel width of 244 m (800 ft) and an overburden thickness of 610 m (2,000 ft). The locations of numbers in the LaModel column represent average results for that layer interval, where the interval is in feet.

Peak Stress and Location

Figure 6 shows the vertical stress profile on the coal in the first 46 m (150 ft) from the edge of the abutment for two generic model cases. Results from LaModel were calculated with 1.5-m (5-ft) elements with eight yield rings, while the results from MULSIM/NL were calculated with 3.0-m (10-ft) elements and four yield rings. Results from FLAC were calculated using 0.76-m (2.5-ft) elements. While these element sizes are not the same, they represent the best results one can get for a single panel model within the constraints of each numerical tool.

The peak stresses and their locations are significantly affected by element size. To adequately compare results using the Mark-Bieniawski formula (MULSIM/NL and LaModel) and an equivalent-strength, elastic-perfectly-plastic constitutive law (FLAC), the elements would need to be the same size and probably smaller (0.76-m (2.5-ft) or smaller).

For cases where only the seam is extracted (no immediate roof shale caves (e.g., Figure 6*A*)), FLAC tends to calculate higher peak stress than MULSIM/NL and LaModel. However, FLAC, MULSIM/NL, and LaModel with layer thickness set at overburden height calculate similar peak stresses and stress profiles.

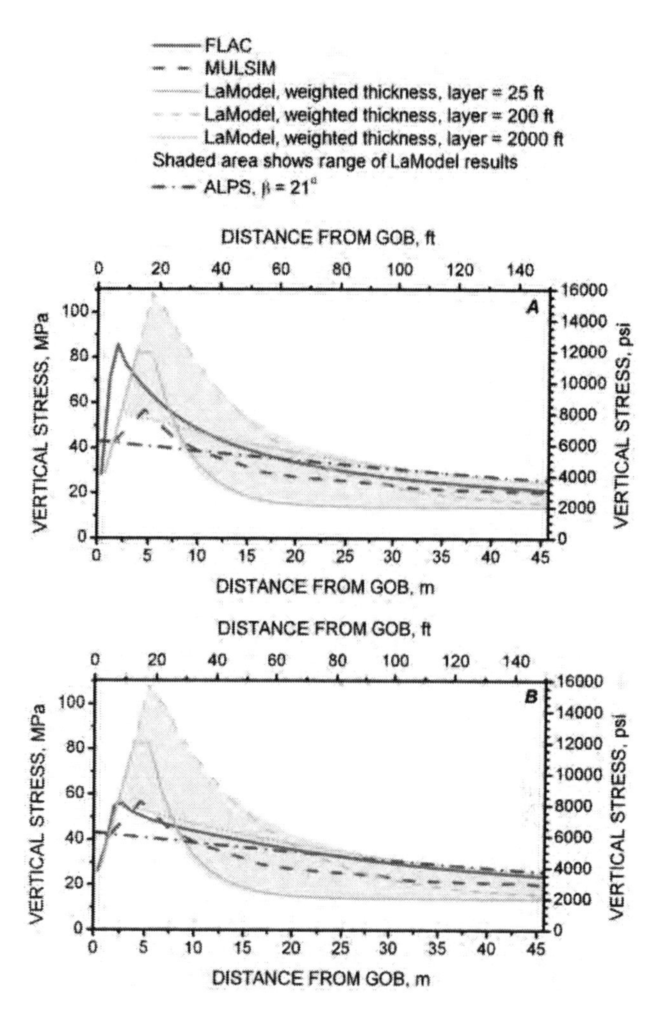

Figure 6.—Vertical stress profile on abutment for the generic model case of 61 m (200 ft) of roof sandstone. Gob is modeled. *A*, No immediate roof shale; *B*, immediate roof shale thickness is 15 m (50 ft).

FLAC, MULSIM/NL, and LaModel handle material failure near the rib differently. In FLAC, zones stressed beyond the elastic limit undergo plastic flow, thus reducing the stiffness near the rib and shifting overburden weight to stiffer coal farther from the rib. MULSIM/NL and LaModel impose an elastic limit according to the Mark-Bieniawski formula and allow more closure of the seam in an element if stresses are above the limit. These two methods are not equivalent and thus can affect the amount of peak stress calculated. It is, therefore, not surprising when FLAC peak stresses are significantly different from either MULSIM/NL or LaModel. ALPS, of course, has no ability to simulate failure of seam material near a rib. Where it is assumed that immediate roof shale caves and forms gob (FLAC results in Figure 6*B*), the location of the top of the seam with respect to the geometry of the opening affects the amount of peak stress. Such geometry is not possible in the boundary-element codes, where full-height extraction of the seam only is assumed.

Stress Transfer To Abutment

Figures 7 and 8 show the range of locations in the abutment where total vertical stress returns to 150% and 200%, respectively, of premining vertical stress for reasonable ranges of input for each tool. It is, in essence, a map of accessible solutions. LaModel lamination thickness significantly affects stress transfer distance. Results with lamination thickness of 7.6-m (25-ft) plot at the bottom of the LaModel range, while results with lamination thickness of 610-m (2,000-ft) plot at the top of the LaModel range. MULSIM/NL stress transfer distance seems to be at approximately midrange of the LaModel results. MULSIM/NL stress does not seem to transfer as far into the abutment as ALPS, FLAC, or LaModel with layer thickness set at the overburden thickness. However, FLAC results with smaller roof sandstone thickness plots near corresponding MULSIM/NL results.

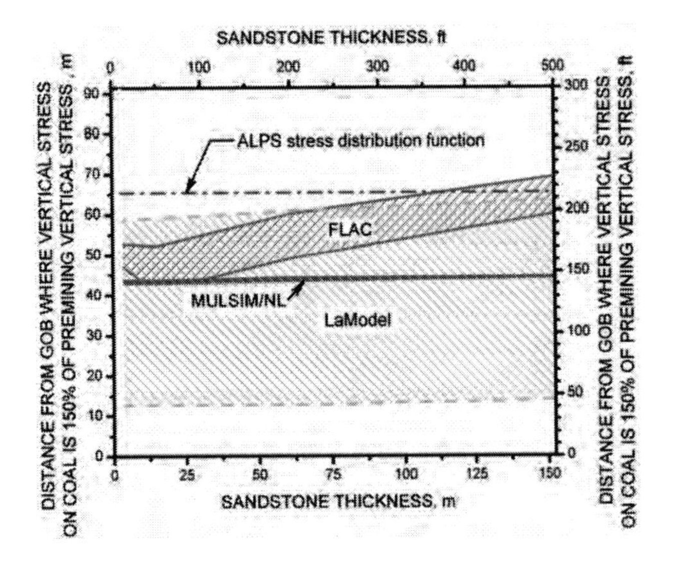

Figure 7.—Distance from gob where vertical stress on coal is 150% of premining vertical stress. Results include immediate shale thickness of 0, 3.0, and 15 m (0, 10, and 50 ft) and all LaModel layer intervals modeled.

Equivalency of MULSIM/NL and LaModel: Generic Model Results

Equivalence of MULSIM/NL and LaModel tools was also tested for the generic model to see if some combination of input parameters might produce essentially equivalent stress distributions. This was addressed by applying generic model *E* and varying *t* in LaModel.

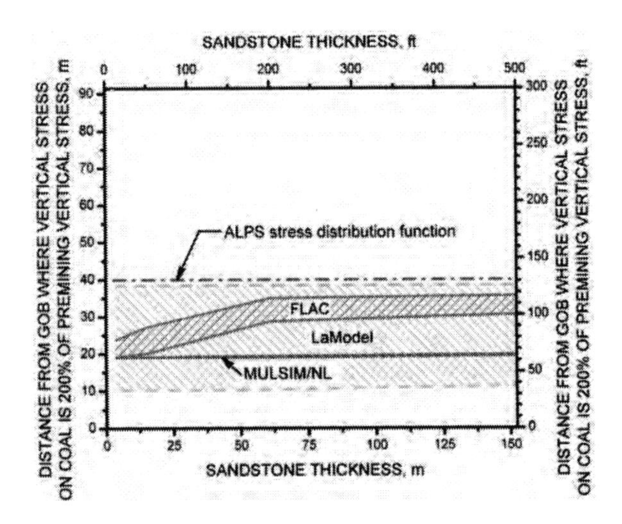

Figure 8.—Distance from gob where vertical stress on coal is 200% of premining vertical stress. Results include immediate shale thickness of 0, 3.0, and 15 m (0, 10, and 50 ft) and all LaModel layer intervals modeled.

Figure 9 shows stress profiles for the generic case where immediate roof shale thickness is 0 m and roof sandstone thickness is 61 m (200 ft). Stress profile equivalency does not seem possible for this practical example with inelastic rib properties. When layer thickness was set to overburden thickness, peak stress calculated by the two tools was nearly the same. However, LaModel results showed more load closer to the opening than those of MULSIM/NL. It seems that no value of the constant *tE* will produce a LaModel stress profile equivalent to that from MULSIM/NL.

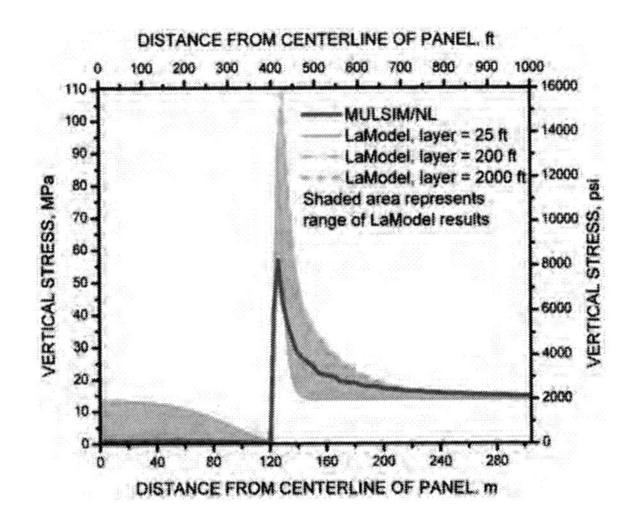

Figure 9.—Vertical stress profiles across half-width of panel and abutment for MULSIM/NL and LaModel for the same overburden properties and various layer thicknesses for LaModel. Properties were determined by the weighted thickness method for shale = 0 m and sandstone = 61 m (200 ft).

Discussion

These results show that the underlying assumptions of each of these stress analysis tools (ALPS, MULSIM/NL, LaModel, and FLAC) significantly influenced results. Of these, FLAC and MULSIM/NL with averaged overburden properties provide the most similar results. The main difference is that FLAC transfers overpanel weight slightly farther into the abutment. The recommended $\beta = 21°$ for ALPS, which controls the portion of overpanel weight transferred to the abutment, underestimates abutment load transfer relative to FLAC and MULSIM/NL. Larger values of β provide closer results, but nonstandard input to ALPS makes comparison with the ALPS database invalid.

LaModel produces the widest range of possible solutions and thus is most sensitive to input parameters. Heasley [2008b] addresses this by recommending a specific input development process. However, in deep western conditions where the stratigraphic column includes a strong, stiff member, overpanel weight may be transferred farther [Barron 1990; DeMarco et al. 1995; Gilbride and Hardy 2004; Goodrich et al. 1999; Kelly 1999; Maleki 2006] than Equation 2 would indicate. In such a case, observations or measurements of load transfer would be necessary to determine the maximum load transfer distance.

In this study, the gob model was not varied. However, accurate simulation of gob behavior is perhaps the input that most significantly affects stress transfer to the abutment. All four tools may be able to adequately simulate site-specific gob behavior where detailed observations or measurements have been made of strata behavior, subsidence, extent of cave, gob loading, etc. However, not all tools can simulate important mechanisms. For example, a strong, stiff overburden member may result in arching of stress over the excavated panel and subsequent sudden collapse. Displacement-discontinuity models cannot simulate that mechanism or its effects. Only models that simulate the behavior of individual stratigraphic members can simulate and/or predict this mechanism.

When moving between MULSIM/NL and LaModel, it may be possible to get equivalent closures at midpanel by limiting error resulting from edge and element size effects, but equivalent stress profiles cannot be achieved in this case. Equivalence seems to be approached, but not attained, as lamination thickness is increased to overburden depth. The user's choice between these two tools should be motivated by measured behavior or characteristics of the stratigraphic column, such as the strength of beds, bedding plane properties, thickness of beds, etc.

CALIBRATION TO STRESS MEASUREMENTS

Comparison to actual conditions is the only way to evaluate the validity of a model. It can also be used to "calibrate" model input, i.e., results of the previous section show only relative performance, not which is "best." Figures 7 and 8 show ranges of stress transfer distance that each tool can achieve for a given geology. Therefore, it is best, where possible, to assess whether a tool is appropriate for a specific site.

As an example, Larson and Whyatt [2009] compared stress in the abutment calculated with ALPS, LaModel, and FLAC with borehole pressure cell measurements of stress induced by mining at two sites. Figures 10 and 11 show the same results with MULSIM/NL calculations added. Gate road entries were not included in the models for convenience. However, aside from local perturbations around entries, each tool can be optimized to provide the best possible approximation of the measured stress distribution.

While the borehole pressure cells were not located close enough to the rib to capture the actual peak stress, the ALPS stress distributions do not simulate the rapid change in stress near the gob. Calibrated MULSIM/NL stress change profiles reasonably matched the changes in stress measurements. LaModel stress profiles varied greatly, with either the peak stress too high or the stress not decreasing as quickly as the measurements with distance from the gob. FLAC models were not built to closely simulate stratigraphy because stratigraphic members or their properties was not sufficiently described [Barron 1990; Koehler et al. 1996]. Instead, generic model geometries with actual panel widths and overburden heights were used to approximate actual site conditions. In these cases, calculated vertical stress profiles with roof sandstone thicknesses of 3 m (10 ft) and 15 m (50 ft) are reasonably close to the changes in stress measurements, suggesting that calculated results likely would match stress change measurements closely if the actual stratigraphic columns were modeled.

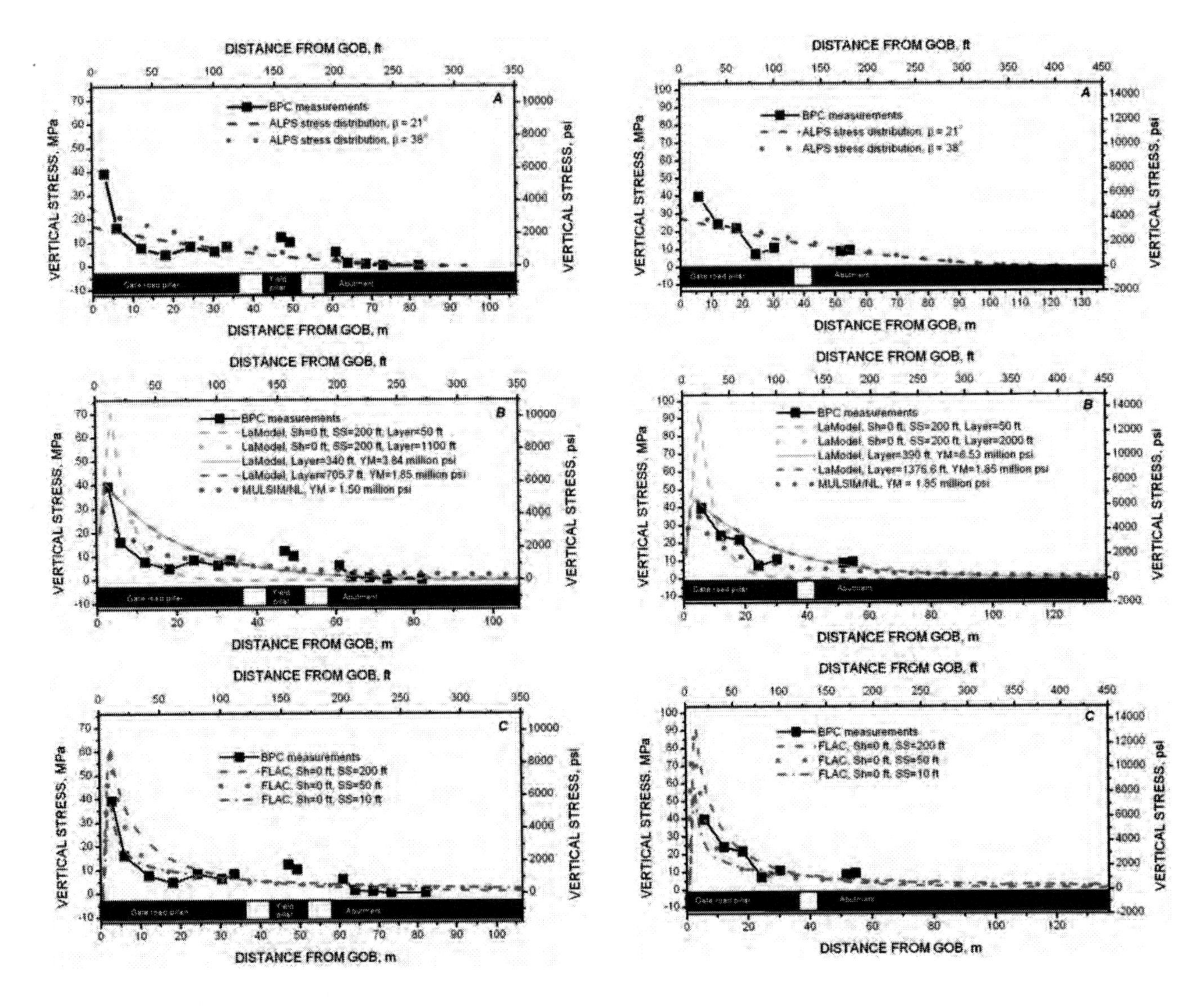

Figure 10.—Mining-induced stress around the 6th Right gate roads at a Utah Mine [Koehler et al. 1996]. *A*, ALPS-assumed stress distribution function and measurements; *B*, MULSIM/NL and LaModel results; *C*, FLAC results with measurements. (BPC = borehole pressure cell; Sh = shale; SS = sandstone; YM = Young's modulus.)

Figure 11.—Mining-induced stress around the 9th East gate roads at a Utah Mine [Barron 1990]. *A*, ALPS-assumed stress distribution function and measurements; *B*, MULSIM/NL and LaModel results; *C*, FLAC results with measurements. (BPC = borehole pressure cell; Sh = shale; SS = sandstone; YM = Young's modulus.)

The importance of calibrating tool input properties to site conditions is emphasized by these results. This point was also underscored in a recent Program Information Bulletin from the Mine Safety and Health Administration [Skiles and Stricklin 2009]. In the bulletin, an eight-step general process is outlined for successful calibration to specific sites. The bulletin states, in summary:

"Successful numerical simulation requires a substantial effort including the observation of in-mine conditions in many areas and the often repetitive process of calibrating model parameters...It cannot be over-emphasized, however, that in order to be of value, a numerical model must be validated and provide a realistic representation of the underground environment for which it is applied."

CONCLUSIONS

Empirical, boundary-element, and volume-element tools are often used to evaluate mine plans. These tools are not interchangeable. Site-specific information must be used to calibrate and assess the appropriateness of each tool. For instance, LaModel does not consider tectonic stress. Thus, it might be a poor choice if high horizontal stresses are present and can influence outcomes in the mine. For empirical tools, application site conditions should be compared with conditions at underlying cases to determine whether the tool might be justified. The published database includes a wide variety of geologic and stress conditions, but it is possible that cases exist outside of database conditions. If possible, additional cases based on local experience should be compared to the published database to establish validity.

Generally, a tool's input properties should be calibrated to site-specific conditions and observations. This point is even more important when a strong, stiff member is present in the stratigraphic column. Calibration procedures should include achieving the correct abutment stress profile and reasonable stress transfer distance. Empirical tools are often an exception, as the reference database typically includes a variety of site conditions. However, if local behavior or cases depart significantly from the underlying database, the user may need to make adjustments. These can include revising coal strength and cave angle, for example. Alternatively, a revised critical stability factor might be proposed. In either case, adjustments delink results from the underlying empirical database and the established success criterion, creating, in essence, a new empirical method that must be justified on its merits.

Results of the generic model study show that assumptions and features of the tools studied differ too markedly for a model constructed with one tool to be "converted" into another through application of the same input parameters. Thus, care must be taken when taking input parameters from past analyses using different modeling programs. New models should be calibrated to field observations and, ideally, measurements of critical behavior.

Results from this study show that if FLAC input accounts for individual stratigraphic members and their properties, then the seam stress profile that it calculates is most similar to results calculated by MULSIM/NL, where overburden properties were determined by averaging individual member properties by some reasonable method. However, the user should evaluate the stratigraphic column and the properties of its individual members with respect to the caving mechanism of the site. An analysis tool should be able to simulate the most important effects of that mechanism.

LaModel results are highly sensitive to input parameters and, therefore, careful model calibration, adjusting tE to fit observed or measured stress conditions, is required. LaModel results did not fit the example cases very well. The quick decrease of stress near the rib requires tE to be relatively small, but this increases peak stress unrealistically.

ALPS stress distribution significantly underestimates peak stress. However, using the default value of β may, in many cases, be a reasonable estimate of total load transferred to the abutment.

In this study, we assumed a model for gob (shale from Pappas and Mark [1993]) and kept that constitutive model constant for all models. If a stiffer gob were used, less stress would be transferred to the abutment. Determining the correct gob stiffness relationship is one of the most important parts of model calibration because that relationship is the most significant factor in determining amount of load transfer to the abutment.

The user must be very careful when selecting the proper analysis tool. If caving behavior can be affected significantly by variability in the strength and stiffness of stratigraphic column members, then a tool should be used that can take that influence into account. This means that tools that assume a single off-seam material that is elastic may miss details of off-seam strata behavior that may be significant.

ACKNOWLEDGMENTS

The authors would like to thank Keith A. Heasley, Ph.D., West Virginia University, Department of Mining Engineering, for discussions that strengthened this paper.

REFERENCES

Agapito JFT, Goodrich RR, Moon M [1997]. Dealing with coal bursts at Deer Creek. Min Eng *49*(7):31–37.

Barron LR [1990]. Longwall stability analysis of a deep, bump-prone western coal mine: case study. In: Peng SS, ed. Proceedings of the Ninth International Conference on Ground Control in Mining. Morgantown, WV: West Virginia University, pp. 142–149.

Beckett LA, Madrid RS [1986]. Practical application of MULSIM/BM for improved mine design. In: Proceedings of the Third Conference on the Use of Computers in the Coal Industry (Morgantown, WV, July 28–30, 1986), pp. 209–219.

Beckett LA, Madrid RS [1988]. MULSIM/BM: a structural analysis computer program for mine design. Denver, CO: U.S. Department of the Interior, Bureau of Mines, IC 9168. NTIS No. PB 88-237565.

Crouch SL, Fairhurst C [1973]. The mechanics of coal mine bumps and the interaction between coal pillars, mine roof, and floor. Minneapolis, MN: University of Minnesota, Department of Civil and Mineral Engineering. U.S. Bureau of Mines contract No. H0101778. NTIS No. PB 222 898.

Crouch SL, Starfield AM [1983]. Boundary element methods in solid mechanics. London: George Allen & Unwin.

Dahlquist G, Björck Å [1974]. Numerical methods. Englewood Cliffs, NJ: Prentice-Hall.

DeMarco MJ, Koehler JR, Maleki H [1995]. Gate road design considerations for mitigation of coal bumps in western U.S. longwall operations. In: Maleki H, Wopat PF, Repsher RC, Tuchman RJ, eds. Proceedings: Mechanics and Mitigation of Violent Failure in Coal and Hard-Rock Mines. Spokane, WA: U.S. Department of the Interior, Bureau of Mines, SP 01-95, pp. 141–165. NTIS No. PB95-211967.

Donato DA [1992]. MULSIM/PC: a personal computer-based structural analysis program for mine design in deep tabular deposits. Denver, CO: U.S. Department of the Interior, Bureau of Mines, IC 9325.

Gates RA, Gauna M, Morley TA, O'Donnell JR Jr., Smith GE, Watkins TR, Weaver CA, Zelanko JC [2008]. Report of investigation: underground coal mine, fatal underground coal burst accidents, August 6 and 16, 2007, Crandall Canyon mine, Genwal Resources, Inc., Huntington, Emery County, Utah, ID No. 42-01715. Arlington, VA: U.S. Department of Labor, Mine Safety and Health Administration.

Gilbride LJ, Hardy MP [2004]. Interpanel barriers for deep western U.S. longwall mining. In: Peng SS, Mark C, Finfinger GL, Tadolini SC, Heasley KA, Khair AW, eds. Proceedings of the 23rd International Conference on Ground Control in Mining. Morgantown, WV: West Virginia University, pp. 35–41.

Goodrich RR, Agapito JFT, Pollastro C, LaFrentz L, Fleck K [1999]. Long load transfer distances at the Deer Creek mine. In: Amadei B, Kranz RL, Scott GA, Smeallie PH, eds. Rock Mechanics for Industry. Proceedings of the 37th U.S. Rock Mechanics Symposium (Vail, CO, June 6–9, 1999). Rotterdam, Netherlands: A. A. Balkema, pp. 517–523.

Haramy K, Magers JA, McDonnell JP [1988]. Mining under strong roof. In: Peng SS, ed. Proceedings of the Seventh International Conference on Ground Control in Mining. Morgantown, WV: West Virginia University, pp. 179–194.

Heasley KA [1998]. Numerical modeling of coal mines with a laminated displacement-discontinuity code [Dissertation]. Golden, CO: Colorado School of Mines, Department of Mining and Earth Systems Engineering.

Heasley KA [2008a]. Back analysis of the Crandall Canyon mine using the LaModel program. Appendix S in: Gates RA, Gauna M, Morley TA, O'Donnell JR Jr., Smith GE, Watkins TR, Weaver CA, Zelanko JC. Report of investigation: underground coal mine, fatal underground coal burst accidents, August 6 and 16, 2007, Crandall Canyon mine, Genwal Resources, Inc., Huntington, Emery County, Utah, ID No. 42-01715. Arlington, VA: U.S. Department of Labor, Mine Safety and Health Administration.

Heasley KA [2008b]. Some thoughts on calibrating LaModel. In: Peng SS, Tadolini SC, Mark C, Finfinger GL, Heasley KA, Khair AW, Luo Y, eds. Proceedings of the 27th International Conference on Ground Control in Mining. Morgantown, WV: West Virginia University, pp. 7–13.

Heasley KA [2009]. Personal communication, March 17.

Jones RE, Pariseau WG, Payne V, Takenaka G [1990]. Sandstone escarpment stability in vicinity of longwall mining operations. In: Hustrulid WA, Johnson GA, eds. Rock Mechanics Contributions and Challenges: Proceedings of the 31st U.S. Symposium (Golden, CO, June 18–20, 1990). Rotterdam, Netherlands: A. A. Balkema, pp. 555–562.

Kelly M [1999]. 3D aspects of longwall geomechanics. In: Ground behaviour and longwall faces and its effect on mining. Exploration and Mining Report 560F, ACARP project C5017.

Koehler JR, DeMarco MJ, Marshall RJ, Fielder J [1996]. Performance evaluation of a cable bolted yield-abutment gate road system at the Crandall Canyon No. 1 mine, Genwal Resources, Inc., Huntington, Utah. In: Ozdemir L, Hanna K, Haramy KY, Peng S, eds. Proceedings of the 15th International Conference on Ground Control in Mining. Golden, CO: Colorado School of Mines, pp. 477–495.

Larson MK, Whyatt JK [2009]. Critical review of numerical stress analysis tools for deep coal longwall panels under strong strata. SME preprint 09-011. Littleton, CO: Society for Mining, Metallurgy, and Exploration, Inc.

Maleki H [1995]. An analysis of violent failure in U.S. coal mines: case studies. In: Maleki H, Wopat PF, Repsher RC, Tuchman RJ, eds. Proceedings: Mechanics and Mitigation of Violent Failure in Coal and Hard-Rock Mines. Spokane, WA: U.S. Department of the Interior, Bureau of Mines, SP 01-95, pp. 5–25. NTIS No. PB95-211967.

Maleki H [1988]. Ground response to longwall mining: a case study of two-entry yield pillar evolution in weak rock. Colo Sch Mines Q *83*(30):1–60.

Maleki H [2006]. Caving, load transfer, and mine design in western U.S. mines. In: Yale DP, Holtz SC, Breeds C, Ozbay U, eds. Proceedings of the 41st U.S. Rock Mechanics Symposium (Golden, CO, June 17–21, 2006). Alexandria, VA: American Rock Mechanics Association.

Maleki H, Agapito JFT, Moon M [1988]. In-situ pillar strength determination for two-entry longwall gates. In: Peng SS, ed. Proceedings of the Seventh International Conference on Ground Control in Mining. Morgantown, WV: West Virginia University, pp. 10–19.

Mark C [1987]. Analysis of longwall pillar stability [Dissertation]. University Park, PA: The Pennsylvania State University.

Mark C [1990]. Pillar design methods for longwall mining. Pittsburgh, PA: U.S. Department of the Interior, Bureau of Mines, IC 9247. NTIS No. PB 90-222449.

Mark C, Chase FE [1997]. Analysis of retreat mining pillar stability (ARMPS). In: Mark C, Tuchman RJ, eds. Proceedings: New technology for ground control in retreat mining. Pittsburgh, PA: U.S. Department of Health and Human Services, Centers for Disease Control and Prevention, National Institute for Occupational Safety and Health, IC 9446, pp. 17–34.

Mark C, Chase FE, Molinda GM [1994]. Design of longwall gate entry systems using roof classification. In: Mark C, Tuchman RJ, Repsher RC, Simon CL, eds. New Technology for Longwall Ground Control. Proceedings: U.S. Bureau of Mines Technology Transfer Seminar. Pittsburgh, PA: U.S. Department of the Interior, Bureau of Mines, SP 01–94, pp. 5–17. NTIS No. PB95-188421.

Molinda GM, Mark C [1994]. Coal mine roof rating (CMRR): a practical rock mass classification for coal mines. Pittsburgh, PA: U.S. Department of the Interior, Bureau of Mines, IC 9387. NTIS No. PB94-160041.

Pappas DM, Mark C [1993]. Behavior of simulated longwall gob material. Pittsburgh, PA: U.S. Department of the Interior, Bureau of Mines, RI 9458. NTIS No. PB93-198034.

Pariseau WG [2007]. Finite element analysis of inter-panel barrier pillar width at the Aberdeen (Tower) mine. Department of Mining Engineering, University of Utah, and Bureau of Land Management, Salt Lake City, UT.

Salamon MDG [1966]. Reconsolidation of caved areas. Transvaal and Orange Free State Chamber of Mines Research Organization, project No. 801/66, research report No. 58/66.

Salamon MDG [1991]. Deformation of stratified rock masses: a laminated model. J S Afr Inst Min Metall *91*(1): 9–26.

Segerlind LJ [1976]. Applied finite element analysis. New York: John Wiley & Sons, Inc.

Sinha KP [1979]. Displacement discontinuity technique for analyzing stress and displacements due to mining in seam deposits [Dissertation]. University of Minnesota.

Skiles ME, Stricklin KG [2009]. General guidelines for the use of numerical modeling to evaluate ground control aspects of proposed coal mining plans. Arlington, VA: Mine Safety and Health Administration, Program Information Bulletin No. P09-03, March 16, 2009. Available at: http://www.msha.gov/regs/complian/PIB/2009/pib09-03.pdf

St. John CM [1978]. EXPAREA: a computer code for analyses of test scale underground excavations for disposal of radioactive waste in bedded salt deposits. St. Paul, MN: University of Minnesota, Department of Civil and Mineral Engineering. DOE contract No. W-7405-ENG-26, report No. Y/OWI/SUB-7118/2.

Turner MJ, Clough RW, Martin HC, Topp LJ [1956]. Stiffness and deflection analysis of complex structures. J Aeronaut Sci *23*(9):805–823, 854.

Zipf RK Jr. [1992a]. MULSIM/NL: application and practitioner's manual. Pittsburgh, PA: U.S. Department of the Interior, Bureau of Mines, IC 9322. NTIS No. PB93-131993.

Zipf RK Jr. [1992b]. MULSIM/NL: theoretical and programmer's manual. Pittsburgh, PA: U.S. Department of the Interior, Bureau of Mines, IC 9321. NTIS No. PB2004-105476.

PRACTICAL APPLICATION OF NUMERICAL MODELING FOR THE STUDY OF SUDDEN FLOOR HEAVE FAILURE MECHANISMS

By Hamid Maleki, Ph.D.,[1] Collin Stewart,[2] Ry Stone,[3] and Jim Abshire[3]

ABSTRACT

As mining continues toward deeper reserves in thick western U.S. coalfields, the control of mining induced-seismicity has become a priority in many operations in Utah, Colorado, and Wyoming. Gradual floor heave, common in many coal mines, historically has not been a safety issue. However, recent occurrences of sudden floor heave accompanied with seismicity in a few deep mines have fueled this investigation. The focus is to identify and study the mechanisms resulting in sudden heaving of the mine floor in western U.S. deep coal mines through analytical calculations and back analyses of an actual event in a cooperative mine.

Based on review of available data in three mines, underground observations, and back analyses of an actual event, we identified the following contributing factors: (1) the presence of stiff stratigraphic units and thick seams mined at depths exceeding 1,000 ft (305 m); (2) the presence of geological discontinuities, reducing the in situ strength of the coal with a calculated factor of safety near 1 for the mine floor; (3) mining approaching areas of higher than normal stress gradient associated with previous mining or structural anomalies or surface topographic highs; and (4) an additional source of energy triggering sudden failure, such as periodic caving or slip along geological discontinuities.

Detailed FLAC3D modeling of the event has improved understanding of the floor heave mechanism. Preliminary examination of seismic data for the actual event in the cooperative mine and underground observations excluded fault slip at the mining horizon as a major triggering mechanism, but highlighted the significance of weak, anisotropic strata conditions near the intersection of two faults. Numerical investigations at the study site point to shear failure of bottom coal during time-dependent strain softening. The model response is not very sensitive to elevated horizontal stress. Although there is no single triggering mechanism for the sudden failure at the study site, we can point to the coincidence of several unfavorable factors, including faulting, stiff stratigraphic units, and altered stress gradients at the multiple-seam crossing and topographic highs.

INTRODUCTION

As mining continues toward deeper reserves in thick western U.S. coalfields, the control of mining induced-seismicity has become a priority in many operations in Utah, Colorado, and Wyoming. Gradual floor heave, common in many coal mines, historically has not been a safety issue. However, recent occurrences of sudden floor heave accompanied with seismicity in a few deep mines have fueled this investigation. The focus is to identify and study the mechanisms resulting in sudden heaving of the mine floor in western U.S. deep coal mines through analytical calculations and back analyses of an actual event in a cooperative mine. Site inspections and mining records from two additional mines were also reviewed and are included in this paper.

This study was initiated in November 2007 to investigate geotechnical factors contributing to floor bumps in a relatively isolated portion of the cooperative mine. The mine is located in the Somerset Coal Basin, near Paonia, CO, where three mining companies have been extracting coal from multiple-seam reserves within the last 2 decades. At the study site, the operator has been extracting coal reserves from the Upper D and Lower and Upper B Seams. After extracting the Upper D Seam reserves in the No. 2 Mine, the operator has been mining in both the Lower and Upper B Seams located approximately 300 ft (91 m) below the D Seam workings.

The mine has employed the longwall mining method to extract the Upper D Seam since 1999 using yield-abutment gate pillar designs while monitoring surface subsidence, collecting detailed geotechnical data in cooperation with MTI staff, and, more recently, monitoring the mining-induced seismicity in cooperation with NIOSH. In the gate roads of these longwall panels, the operator used yield abutment pillars from 98 to 114 ft (30 to 35 m) wide. Yield pillars were 37–47 ft (11–14 m) wide, and the longwall face was 824 ft (251 m) wide. The mine layout and location of the study area are shown in Figure 1, along with historic workings in the U.S. Steel Mine and King Mine. Similar gate pillar layouts are used in the B Seam while offsetting the position of gate roads with respect to D Seam gate roads (Figure 1).

[1]Maleki Technologies, Inc. (MTI), Spokane, WA.
[2]Formerly with Bowie Resources, LLC, Paonia, CO.
[3]Bowie Resources, LLC, Paonia, CO.

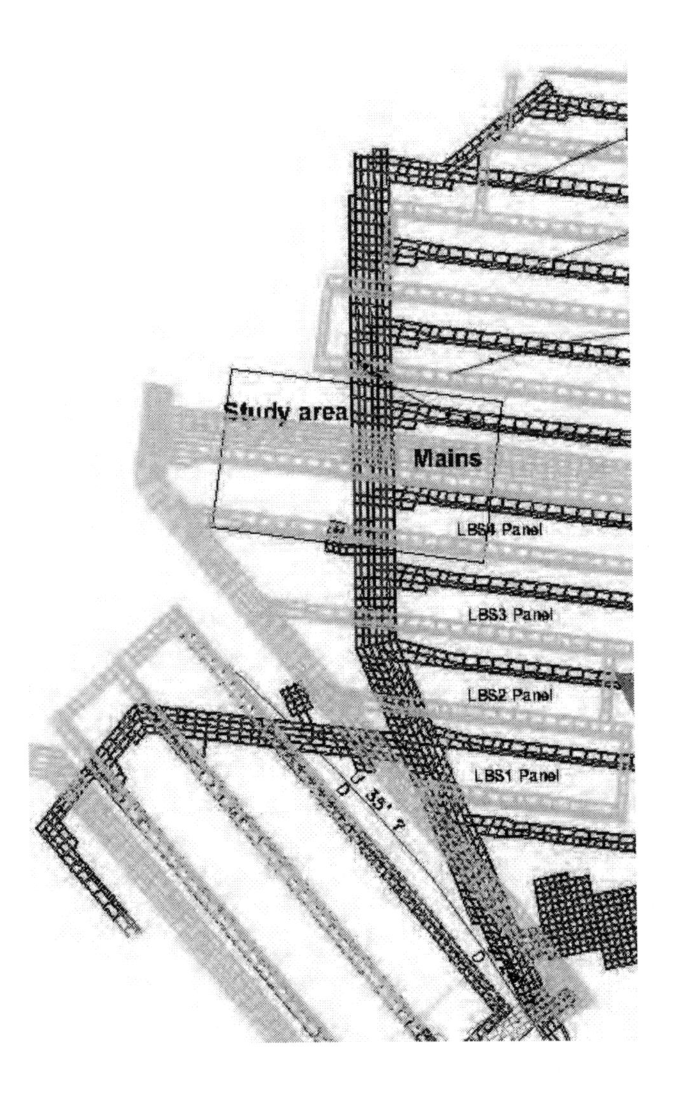

Figure 1.—Mine layout and the location of the study area.

In addition to surface monitoring, the operator implemented a detailed geotechnical program consisting of drilling continuous coreholes from the surface, depositional and structural logging of the core, geologic mapping, rock mechanics testing, and hydrologic monitoring in cooperation with MTI staff at one location. Overburden characteristics were evaluated using Bieniawski's Rock Mass Rating system, identifying massive overburden units capable of transferring loads and reducing differential movements for future extraction of the B Seam [Maleki et al. 2007].

Additional data have been systematically collected near the longwall block using interseam core drilling and the access from the Upper D Seam, exploring geologic and rock strength of the interburden rocks. These activities included lithologic and structural logging, together with mechanical property testing of near-seam strata.

GEOLOGIC SETTING

The Somerset Coalfield is located on the southeastern margin of the Piceance Basin, which lies north of the Gunnison Uplift, west of the Elk Mountains, east of the Uncompahgre Uplift, and south of the White River Uplift. There are six primary coal seams within the Somerset Coalfield ranging from 10 ft (3 m) thick on a single bench of coal to over 20 ft (6.1 m) where one or more seams merge [Maleki et al. 1997]. An estimated 1.5 billion tons of bituminous coal lies within the Somerset field. Coal seams B, C, D, E, and F have been mined historically in this basin. The coal seams contain a very well developed cleat set oriented at N 58° to 75° E [Carroll 2003].

The regional structural dip of the Mesa Verde Formation is 3° to 5° to the north-northeast. The predominant jointing of sedimentary rocks is N 68° E to N 74° E in the basin, with secondary jointing at N 18° W to N 35° W. Regional joint patterns have been mapped by Carroll [2003] and the authors.

During extraction of the D1–D9 and WD1–WD3 panels in the southwest district, we studied faulting using both surface and underground mapping by the geologic staff. Both normal and strike-slip faults have been encountered underground, but only one fault has surface expression. The southern portion of the D6 panel and part of D5 gate road is traversed by D-5 fault, a N 80° to 90° E left-lateral, strike-slip fault system. The system is a complex array of anastomosing fault planes of variable offsets. Relative motion is 10° to 15° from the horizontal with 5–10 ft (1.5–3.0 m) of total throw down to the south. North of the strike-slip fault, the D6–D9 panels are intersected by several N 60° E normal faults with limited lateral extent and displacements appearing as tensile gashes. It seems that these faults formed as a result of differential movement along the strike-slip fault. This assertion is based on observations that throw decreases with increasing distance from the fault to the north, but no such features are found to the south. Where present, kinematic indicators suggest dip-slip movement and little to no shear [Robeck 2005].

Additional data were collected more recently at the location of the floor bump. As shown in Figure 2, geologic structures are projected based on underground mapping in the D Seam as well as available exposures within the B Seam in close proximity to the study area. The east-west oriented D-5 fault shown in Figure 2 has strike-slip movements with very limited vertical offset. As it bends to the north near the event site, the offset increases to 20 ft (6.1 m), becoming near parallel to another set of northwest-oriented faults in the study area. Clearly, the geologic and stress conditions are complex near the event site, and the possibility of abnormal horizontal stress conditions (including reorientation and higher magnitudes) cannot be ruled out. Faulting did not create any significant operational problems or seismicity during multiple crossings at the Upper D Seam. In the weaker, deeper B Seam workings near the bend zone, the faulting has promoted floor heave mostly in a gradual manner, but also suddenly under certain conditions.

Figure 3 shows a typical geologic cross-section for the two-seam longwall area. The major units present are representative of the Cretaceous Age Lower Mesa Verde Formation in this area and include the Rollins Sandstone, the C-Sandstone (locally referred to as Upper Marine Sandstone), and coal seams A through D-Upper. Horizontally bedded mudstone and siltstone sequences, together with the cross-bedded to laminar-bedded fluvial sandstone channels, are also present, but shown as undifferentiated in the cross-section. The alluvial valley fill is believed to be Quaternary in age and is composed of Tertiary Age igneous boulders, cobbles, and gravel.

The coal seams and the Rollins and C-Sandstones are the only truly tabular sedimentary bodies in the section. The two sandstone bodies are even more laterally continuous than the thinner coal seams. Because of strength and lateral persistency, these two marine sandstone bodies have by far the best characteristics for bearing and distributing loads. The C-Sandstone is approximately 90 ft (27 m) thick in two parts with a 10- to 15-ft (3.0- to 4.6-m) mudstone-rich zone near the center of the unit.

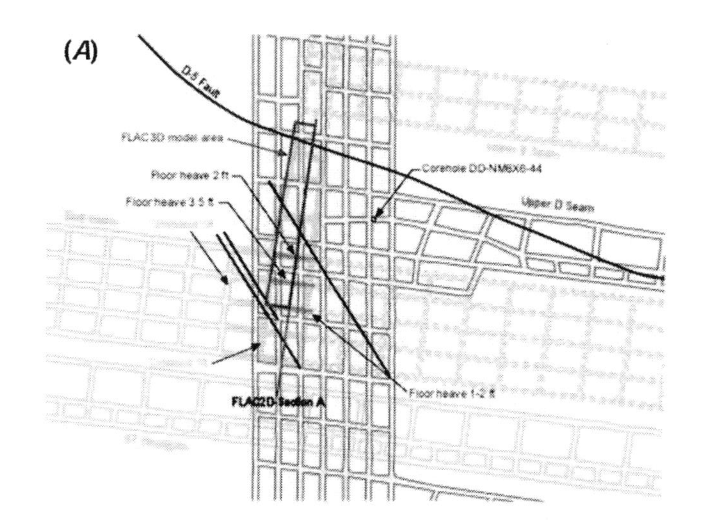

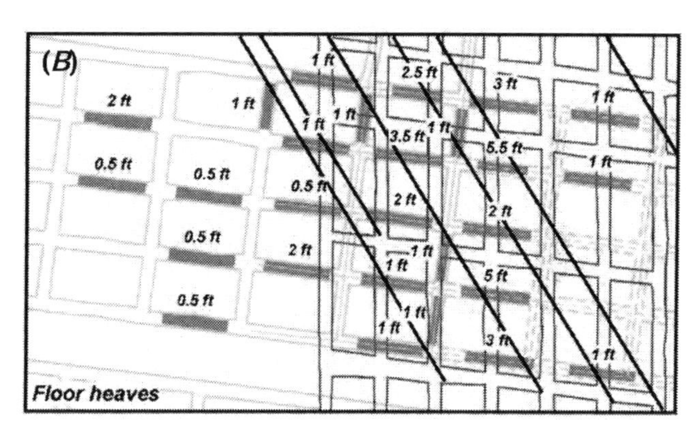

Figure 2.—Mining geometry, geologic conditions near the event locations, and observed floor heave in the Lower B Seam: (*A*) first event, (*B*) final event.

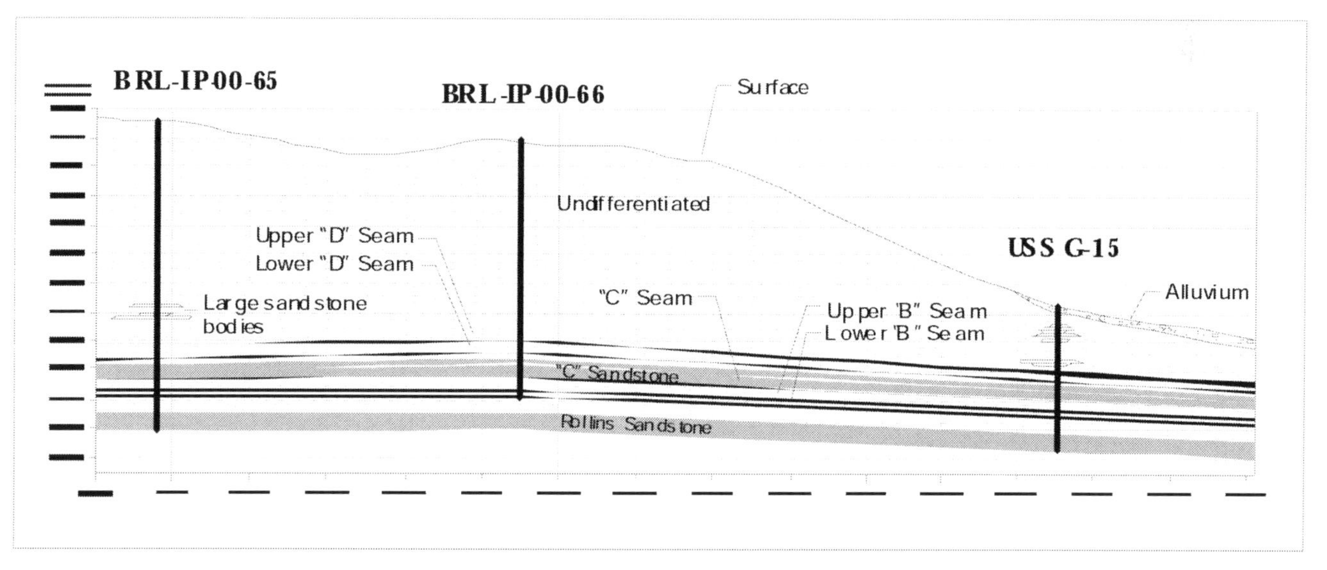

Figure 3.—East-west geologic cross-section.

STRATA STRENGTH AND STABILITY

Figure 4 shows a portion of a lithology and structure log for a corehole drilled from the Upper D Seam toward the B Seams near the event site. Uniaxial compressive strength (in psi) derived from point load tests is also shown in the figure. The data indicate the presence of both weak but mostly strong rocks near the seam, as well as thick sandstone above the Upper B Seam. Using regression analyses of strength and elastic modulus, the majority of rocks are both strong and stiff. These stiff units have the capacity to absorb high stress and release the energy upon failure in the form of bounces or seismic events. Mudstone cap rock failure is common in the study area and is influenced by unfavorable orientation of N 60° E joints with respect to the rooms.

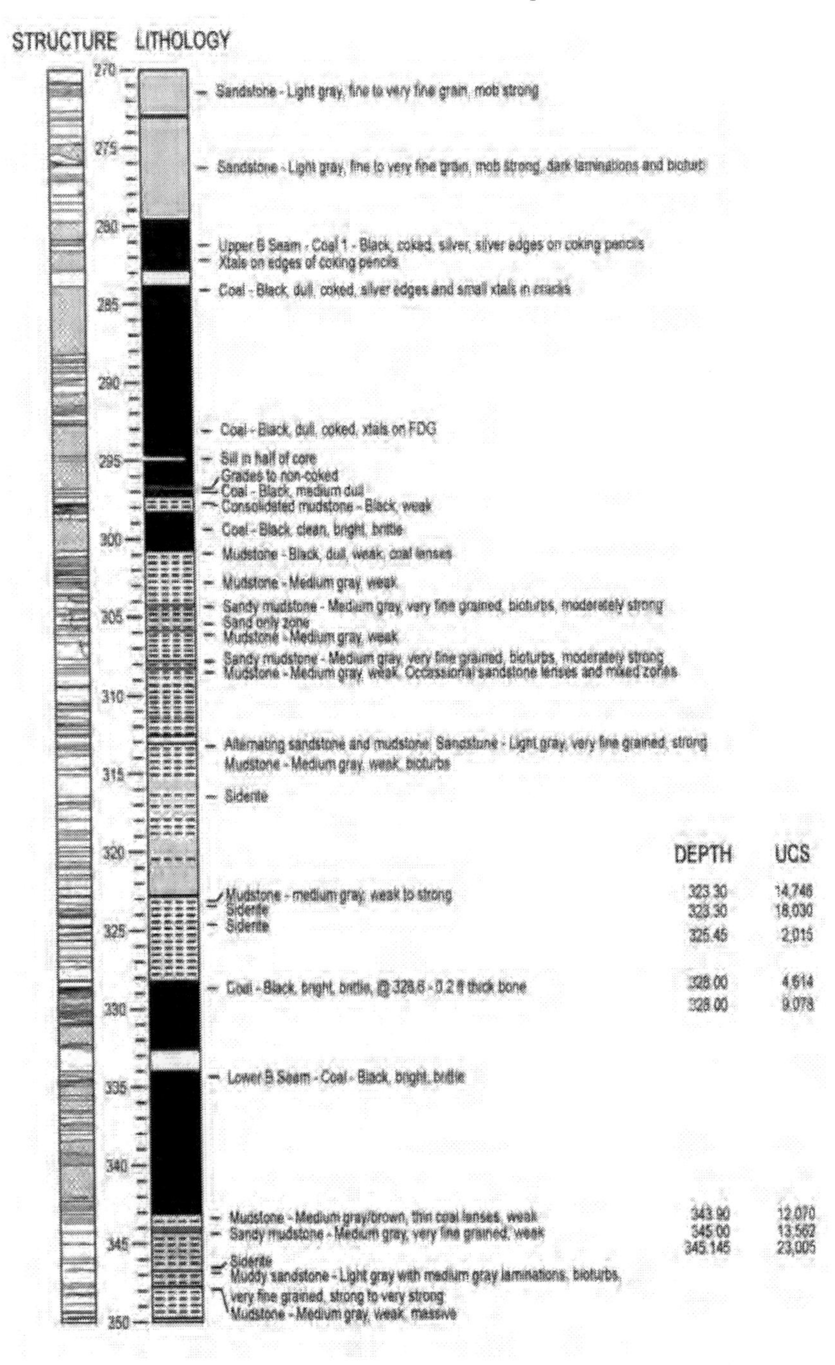

Figure 4.—Lithology and strength log for an interseam corehole close to the event site. (Depth is given in feet. Uniaxial compressive strength (UCS) is given in psi.)

There are many stress measurements in the Somerset Basin that are generally consistent and thus provide a good estimate of far-field horizontal stress. The U.S. Bureau of Mines (USBM) originally measured the far-field horizontal stress at two locations at the neighboring Orchard Valley Mine [Bickel 1993], while the site operator provided another measurement more recently in the No. 2 mine in the D Seam.

Regional measurements show excess horizontal stress [Maleki et al. 1997, 2007] in the Somerset Coalfield along the regional structural trend (N 74° to 80° E). Horizontal stress is only gravity-induced along the secondary direction (N 10° W). The excess stress is moderate, but increases slightly with depth at the study site. The far-field horizontal stress is very anisotropic in this basin at mining depths and away from drainages. The ratio of maximum to minimum secondary principal horizontal stress (P/Q ratio) is approximately 4 to 6. The orientation of maximum principal stress is consistent, averaging N 80° E near the study mine. The stress field may be different at the study area because of the complex structural setting. Maleki et al. [2003, 2006] address an appearance of a switch in P and Q under drainages and other geologic conditions.

The failure occurred on October 2, 2007, accompanied by heave of the bottom coal within three rooms (Figure 2*A*), rapid displacement of coal into these rooms, knocking down of one stopping, and damage to three other stoppings. The failure was accompanied by a seismic event. This event registered 2.1 on the seismic system operated on the surface by the mining company with support from NIOSH. During these initial investigations and modeling, two other events registered on the NIOSH system, which resulted in additional floor heave after resumption of mining activities (Figure 2*B*). The study area in B Seam mains remained bump-free for the last quarter after mining extended beyond the faulted ground. The failure occurred at the vertical intersection of the D and B mains, where the overburden thickness peaked at 1,700 ft (518 m).

Because faulting was significant at the study area, we briefly reviewed potential impacts and mechanisms. Faulting is known to (1) introduce fluids and thus create difficult mining conditions; (2) reduce strata strength; (3) retard load transfer across fault planes, causing localized stress concentrations; (4) change the stiffness of the mine loading system, thus contributing to sudden failure; and (5) influence periodic cave conditions.

The impact of faulting in western U.S. coal mines has been studied through long-term measurements within panel workings and back in the gob [Maleki 1981]. Roof and floor stability problems are likely within fault zones depending on fault characteristics (displacement, fluids, gouge thickness, and composition) and are influenced by lower strata strength and higher moisture content. Because shear stress cannot transmit easily across many faults, localized stress concentration occurs near fault zones, contributing to stability problems during retreat depending on the angle of incidence and mining geometry. Based on pressure measurements in the gob, the interaction of faulting with sudden, periodic caving has been shown to trigger failure of marginally stable structures in the mine roof or floor [Maleki 2006].

POTENTIAL FAILURE MECHANISM

Gradual floor heave historically has not been a safety issue at the cooperating mine and at many other deep western U.S. operations except where it contributed to asymmetrical loading and tilting of cribs [Maleki 1988]. However, recent occurrences of sudden floor heave accompanied with seismicity in a few deep mines have fueled this investigation.

Factors contributing to sudden failure have been studied by Maleki et al. [1997] through analyses of violent failure from 25 case studies in U.S. mines. Those specifically relating to floor heave events are expanded here, including (1) the presence of thick seams and stiff stratigraphic units within the roof and floor; (2) the presence of geological discontinuities, reducing the in situ strength of the bottom coal with marginal stability; (3) mining approaching areas of higher than normal stress gradient associated with previous mining or structural anomalies or surface topographic highs; and (4) an additional source of energy triggering sudden failure, such as periodic caving or slip along geological discontinuities or reaching topographic highs.

Preliminary examination of seismic data for the actual event in the cooperative mine [Swanson 2008] and underground observations excluded significant fault slip at the mining horizon as the triggering mechanism. Although the final release of the energy was sudden, we suspect time-dependent growth of failure around the excavation during the idle time leading to the release of the energy. Such a process is likely to be associated with microseismic emissions (rock noise) and thus provides a means of studying abnormal areas having the potential to deform suddenly. The regional seismic monitoring system operated by NIOSH on the surface became a valuable tool for studying changes in microseismic activities, confirming a history of low-level seismicity at this relatively isolated area at the study site.

Assuming elastic strata conditions, vertical load transfer from the D Seam longwalls are expected to be minimal at a distance of 1,000 ft (305 m) from the edge of the D Seam gob. This distance corresponds to the horizontal distance between the nearest gob and the event location. The influence of any abnormal vertical stress gradient was evaluated at the site using numerical modeling techniques. Other mechanisms relating to diminished ability of the mine loading system [Salamon 1970] to transfer loads in a faulted zone was not considered to be significant at the event site.

Two classic floor heave mechanisms are proposed: (1) buckling failure [Aggson 1978] and (2) shear failure. For the buckling mechanism, Euler's equation shows that the compressive failure of the floor beam is controlled by horizontal stress, beam thickness, and span. This process could be facilitated near the event location by anisotropic strength of strata, higher suspected horizontal stress gradients and reorientation (northwest) near the faults, and laminated nature of the coal in the floor and entry spans. Numerical modeling is used to study the failure mechanisms and examine the significance of these factors.

STRESS ANALYSES

Additional insight is gained by conducting stress analyses to identify significant factors contributing to the event at the study site.

The scope of the modeling was as follows:

1. Preliminary evaluation of the sensitivity of strata deformation to several parameters, including a horizontal stress field and material properties along a north-south oriented section (section A) using a two-dimensional multilayer FLAC model (FLAC2D) (Figure 2*A*). The intent was to reduce the number of potential causal factors in preparation for more detailed analyses.
2. Detailed sensitivity analyses and evaluation of floor heave failure mechanisms with the three-dimensional FLAC analyses (FLAC3D) (Figure 2*A*). Modeling included three different constitutive material models and simulation of distinct faulting.

FLAC2D SENSITIVITY ANALYSES

The main objective of these preliminary analyses was to evaluate the sensitivity of strata deformation to several parameters, including an abnormal horizontal stress field, higher vertical stress at main crossings, and weak material properties as influenced by geologic structure. The modeling is completed along an approximately north-south oriented section (Figure 2*A*) using the two-dimensional multilayer FLAC code (FLAC2D). These are inelastic analyses that considered yielding and redistribution of stress during the mining process using the Mohr-Coulomb criteria.

The modeling scope included preliminary evaluation and ranking of causal factors that were suspected to have played a role at the event site:

- Higher vertical stress gradient with a worst-case assumed scenario of 40% increase.
- Twice higher horizontal stress gradient.
- Twice lower horizontal stress gradient.
- Weaker floor rocks as influenced by faulting and simulated by reducing the cohesion of floor.
- Bedding planes with low shear strength properties at the Lower B Seam contact and 12 ft (3.7 m) below as simulated using the interface model embedded in FLAC.

The far-field horizontal stress has been measured underground with the maximum stress oriented near east-west. Both the orientation and direction of the horizontal stress could change at the fault zone. The impact of both the higher and lower northwest component of stress is evaluated in these simulations.

Results mostly excluded from this paper showed the calculated floor heave to be sensitive to significant reduction in bottom coal cohesion near the fault, the presence of weak bedding planes at the Lower B Seam contact with floor rocks, and simulated high horizontal stress. On the contrary, the model does not predict the heave to be sensitive to significant changes in vertical stress gradient or weak interfaces deeper into the floor (modeled 12 ft (3.7 m) below the floor). For the worst combination of these factors, the calculated floor heave approaches reported values at the first event location (1 ft (0.3 m)). Since many of these significant factors could be associated with northwest-oriented faulting, it is this local additional faulting that seems to have strongly influenced the floor heave events.

Typical safety factor is reported under "strength/stress ratio" in Figure 5 for an analysis with twice higher horizontal stress gradient. The marginal stability of the mine floor and very stable pillar core are noted for the assumed conditions.

FLAC3D ANALYSES

FLAC3D (F3D) modeling advances FLAC2D (F2D) results by including the anisotropic nature of horizontal stress, different material models, as well as three-dimensional excavation geometries. The location of the analyzed region is shown in Figure 2*A*, where model boundaries are lines of symmetry. As with the F2D analyses, we evaluated the impact of higher than normal stress gradient. In addition, the anisotropic strength along either the face cleat or butt cleat were analyzed together with distinct and equivalent modeling of faults. The butt cleat was intensified along the northwest-oriented fault planes. An observation of significance to F3D modeling was the higher pattern of heaving along the rooms and crosscuts, but notably absent at the intersections.

In these analyses, we examined both normal (measured) horizontal stress, with the maximum stress oriented parallel to the rooms, and high stress gradient. The latter assumes an abnormally high horizontal stress field influenced by D-5 and northwest-oriented faults and simulated by applying up to 2.5 times higher horizontal stress parallel to the crosscuts (SYY). In addition, we implemented three different constitutive models for the coal, including Mohr-Coulomb plasticity, Mohr-Coulomb strain softening, and ubiquitous joints.

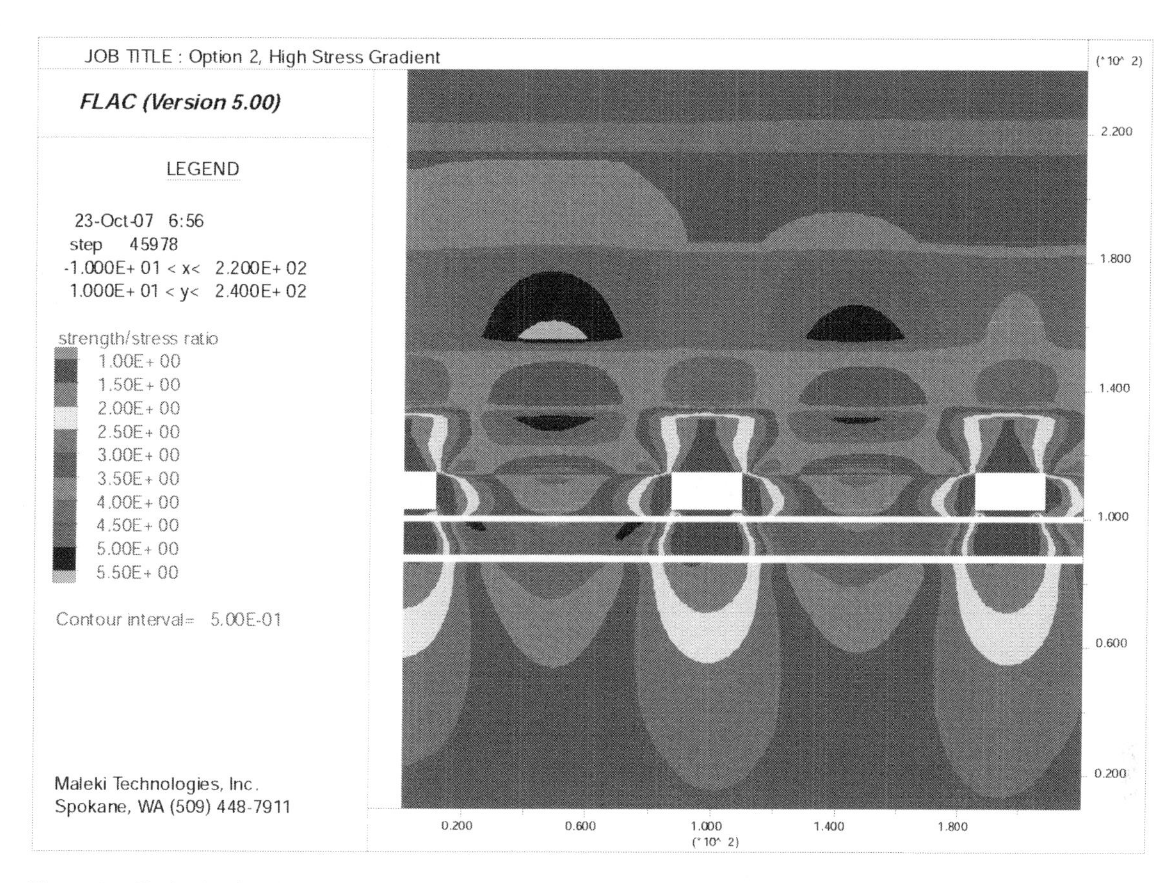

Figure 5.—Typical safety factor plot for high horizontal stress gradient. Note marginal stability of the bottom coal.

The significance of anisotropic strength properties of the coal seam along the face cleat and the butt cleat (paralleling the northwest fault trend) was also evaluated. At the event site, the geologic staff has mapped weaker, slickensided butt cleats at fault planes. In addition, the face cleat is known to influence rib movements in the mine due to its persistence and small deviation from the orientation of rooms (30°). Using the bilinear strain-softening ubiquitous-joint plasticity model embedded in FLAC, we compared the model response to observed weakness planes along the face or butt cleat.

Figure 6 compares calculated floor heave for three constitutive models supporting gradual strain softening near the excavation. Note higher calculated heave at the intersections for the Mohr-Coulomb model, which is replaced by significantly higher heave along the rooms and crosscuts as the strain-softening model is used for the coal and ubiquitous-joint model for the face cleat. By including the weaker nature of the coal along the face cleat and using the strain-softening ubiquitous-joint model, we come to close agreement between observed and measured floor heave patterns. This concurs with another comprehensive case study including detailed measurements of stress, deformation, and floor-bearing capacity [Maleki 1993; Maleki and Hollberg 1995].

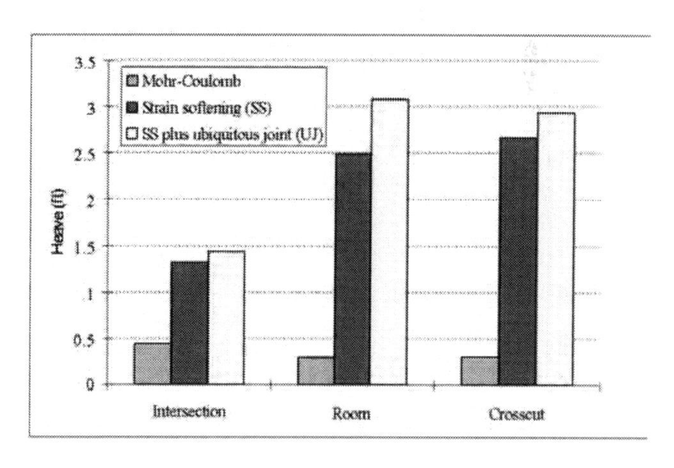

Figure 6.—Comparison of floor heave for different material models.

Figure 7 compares calculated floor heave for four analyses designed to simulate approximate faulting using both distinct and equivalent strain-softening zones. Using an interface to simulate the north-south oriented fault plane, the model closely predicted the pattern of floor heave in the study site. However, the magnitudes were lower because of high-strength values used to overcome model convergence issues. Alternatively, by simulating two north-south oriented equivalent fault zones using the strain-softening model, the model came to replicate significant heave along the rooms near the fault crossings. The higher assumed north-south horizontal stress makes some improvement in predicting floor heave patterns, but the model is rather insensitive to simulated changes in the horizontal stress (Figure 8). This supports strain softening near the excavation and along the fault planes as the primary factor influencing a large amount of observed heave (Figure 9).

The model has helped greatly in understanding floor heave mechanisms and the significance of faulting. With planned additional geotechnical measurements, we intend to investigate the source of sudden heaving at the study site. With no nearby longwall extraction and evidence of slip, noticeable slip along the fault at the mining horizon is excluded. Because model response is not very sensitive to simulated higher than normal horizontal stress gradients, we do not consider high horizontal stress and buckling failure to be likely at the site. Although there is no single triggering mechanism for the sudden failure, we can point to the coincidence of several unfavorable factors, including faulting, stiff stratigraphic units, and altered stress gradients at the multiple-seam crossing and topographic highs.

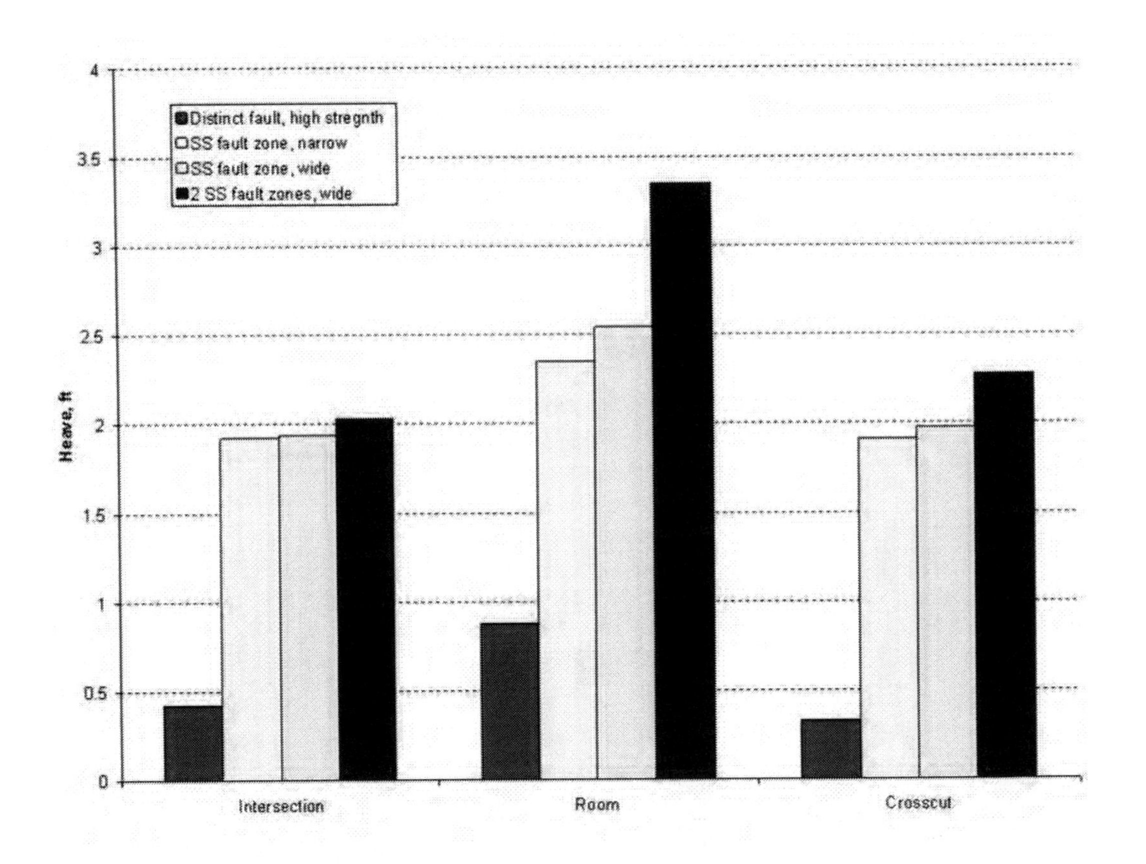

Figure 7.—Calculated floor heave for different methods of simulating faulting in the study area.

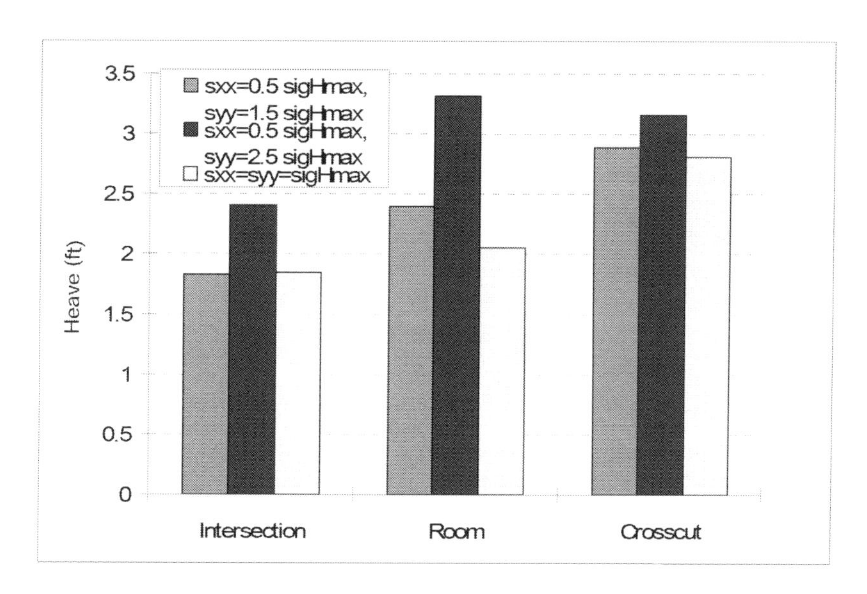

Figure 8.—Calculated floor heave for different horizontal stress gradients.

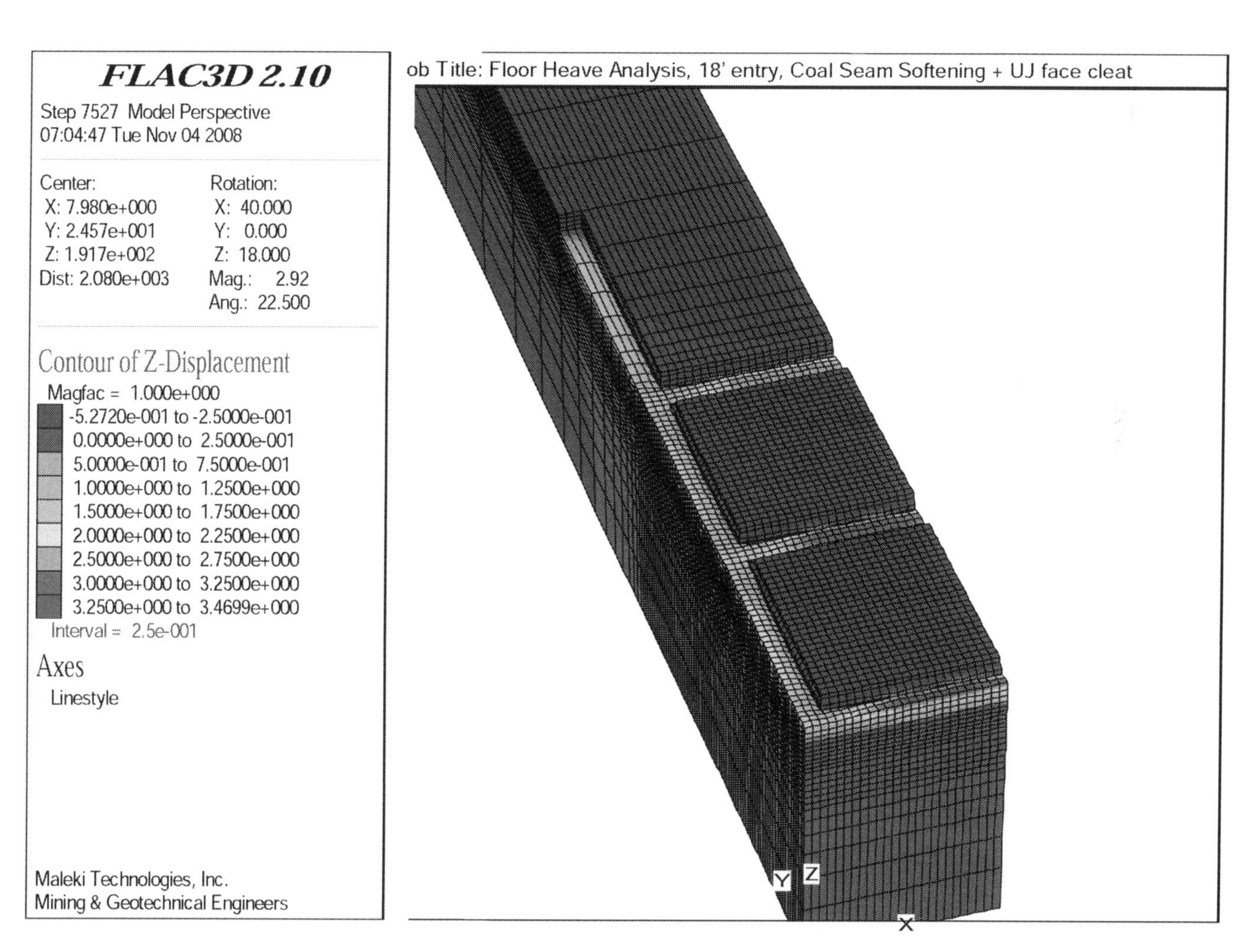

Figure 9.—Vertical displacement contours for equivalent simulated faulting. (UJ = ubiquitous joint.)

COLLECTIVE EVIDENCE FROM OTHER MINES

We also inspected and studied factors contributing to sudden floor heave at two other western U.S. mines at fault crossings. Despite many similarities between geologic setting and rock mass conditions, the sudden floor heave occurred during longwall retreat at those sites where significant changes in stress concentration results about the gob cavity. In addition, nonuniform caving and periodic collapse of overburden rocks have been measured at one of these sites through direct-pressure measurements in the gob [Maleki 1981, 2008]. Such events have adequate energy to trigger sudden floor heave in marginally stable bottom coal in faulted ground with low-strength properties. Because the event happened as the longwall face approached a distance equal to face width, we suspect that the caving provided the additional energy at the mine based on past measurements and regional experiences [Maleki 1981].

CONCLUSIONS

Common factors identified at all sites include (1) the presence of stiff units and thick seams mined at depths exceeding 1,000 ft (305 m); (2) the presence of geological discontinuities, reducing the in situ strength of the coal with a calculated factor of safety near 1 for the mine floor; (3) mining approaching areas of higher than normal stress gradient associated with previous mining or structural anomalies or surface topographic highs; and (4) an additional source of energy triggering sudden failure, such as periodic caving or slip along geological discontinuities.

Preliminary examination of seismic data for the actual event in the cooperative mine and underground observations excluded fault slip at the mining horizon as a major triggering mechanism, but highlighted the significance of weak, anisotropic strata conditions near the intersection of two faults. Numerical investigations at the study site point to shear failure of bottom coal during time-dependent strain softening. The model response is not very sensitive to elevated horizontal stress. Although there is no single triggering mechanism for the sudden failure at the study site, we can point to the coincidence of several unfavorable factors, including faulting, stiff stratigraphic units, and altered stress gradients at the multiple-seam crossing and topographic highs.

It should be emphasized that faulting is often not associated with abnormal far-field stress conditions based on measurements by the USBM and decades of mining experience. However, there are some complex structural settings and stress conditions that are known to be associated with higher than normal stress gradients. The identification and projection of those conditions ahead of mining, together with monitoring of roof-floor convergence, play a key role in predicting zones with higher potential for sudden floor heave.

Practical research should continue toward improving the understanding of sudden failure through site-specific measurements and numerical modeling of other events at cooperating mines.

REFERENCES

Aggson JR [1978]. In-situ stress fields and associated mine roof stability. Mini-Symposium Series No. 78-1, SME-AIME Annual Meeting, Dallas, TX, 7 pp.

Bickel D [1993]. Rock stress determinations from overcoring: an overview. Denver, CO: U.S. Department of the Interior, Bureau of Mines, Bull. 694.

Carroll CJ [2003]. Fractures in the Mesaverde Group at Somerset Coal Field, Delta and Gunnison Counties, Colorado. Denver, CO: Colorado Geological Survey.

Maleki H [1981]. Coal mine ground control [Dissertation]. Golden, CO: Colorado School of Mines, Vol. 1.

Maleki H [1988]. Ground response to longwall mining: a case study of two-entry yield pillar evolution in weak rock. Colo Sch Mines Q *83*(3).

Maleki H [1993]. Case study of the inelastic behavior of strata around mining excavations. In: Haimson B, ed. Proceedings of the 34th U.S. Symposium on Rock Mechanics (Madison, WI, June 27–30, 1993), vol. 1, pp. 137–140.

Maleki H [2006]. Caving, load transfer, seismicity in four Utah mines. In: Peng SS, Mark C, Finfinger GL, Tadolini SC, Khair AW, Heasley KA, Luo Y, eds. Proceedings of the 25th International Conference on Ground Control in Mining. Morgantown, WV: West Virginia University, pp. 79–86.

Maleki H [2008]. In situ pillar strength and overburden stability in U.S. mines. In: Peng SS, Tadolini SC, Mark C, Finfinger GL, Heasley KA, Khair AW, Luo Y, eds. Proceedings of the 27th International Conference on Ground Control in Mining. Morgantown, WV: West Virginia University.

Maleki H, Hollberg K [1995]. Structural stability assessment through measurements. In: Proceedings of the ISRM Workshop on Rock Foundations (Tokyo, Japan).

Maleki H Clanton J, Koontz W, Papp A [1997]. The relationship between far-field horizontal stress and geologic structure. Rock Support, Oslo, Norway.

Maleki H, Dolinar DR, Dubbert J [2003]. Rock mechanics study of lateral destressing for the advance-and-relieve mining method. In: Peng SS, Mark C, Khair AW, Heasley KA, eds. Proceedings of the 22nd International Conference on Ground Control in Mining. Morgantown, WV: West Virginia University, pp. 105–113.

Maleki H, Stewart C, Hunt G [2006]. Subsidence characteristics for Bowie Mines, Colorado. In: Proceedings of the 41st U.S. Rock Mechanics Symposium. Alexandria, VA: American Rock Mechanics Association.

Maleki H, Stewart C, Stone R [2007]. Three-seam stress analyses at Bowie Mines, Colorado. In: Peng SS,

Mark C, Finfinger GL, Tadolini SC, Khair AW, Heasley KA, Luo Y, eds. Proceedings of the 26th International Conference on Ground Control in Mining. Morgantown, WV: West Virginia University, pp. 29–37.

Robeck E [2005]. The effects of fault-induced anisotropy on fracturing, folding, and sill emplacement [Thesis]. Provo, UT: Brigham Young University.

Salamon M [1970]. Stability, instability and design of pillar workings. Int J Rock Mech Min Sci *7*:613–631.

Swanson P [2008]. Monitoring mine seismicity with an automated wireless digital strong-motion network. In: Peng SS, Tadolini SC, Mark C, Finfinger GL, Heasley KA, Khair AW, Luo Y, eds. Proceedings of the 27th International Conference on Ground Control in Mining. Morgantown, WV: West Virginia University, pp. 79–86.

ADVANCED NUMERICAL SOLUTIONS FOR STRATA CONTROL IN MINING

By Axel Studeny[1] and Carsten Scior[2]

ABSTRACT

Geomechanical-numerical methods are an established tool for planning of underground excavations worldwide. By describing the interaction between heterogeneous layered strata and support elements, it is possible to determine the suitability of alternative support systems.

The German hard-coal mining industry has more than 20 years of experience in investigating and using geotechnical parameters to conduct numerical calculations. Today, nearly all excavations (shafts, pit bottoms, bunkers, coal faces, development drivages, gate roads, etc.) are evaluated by geomechanical-numerical tools in combination with empirical planning methods.

A high accuracy of planning results, which is especially required in the German coal industry with its deep-level longwalls, bedded and weak surrounding strata, and multiple-seam mining layouts, depends on the extensive calibration of the numerical models, which was achieved by:

1. Taking a large number of underground measurements with empirical evaluation
2. Carrying out physical modeling
3. Using the characteristic features of the installed support elements

This paper describes the planning approach for numerical models in mining and presents the advantages of these planning tools.

INTRODUCTION

Numerical methods are widely used for solving a wide variety of technical problems. They are mainly used in applications where analytical and empirical processes cannot provide the required degree of accuracy because of complex boundary conditions [Junker 2006].

The mining industry uses geomechanical-numerical models extensively as operational planning tools. Analytical and empirical methods have thus far proved to lack the degree of precision required for describing the interaction that occurs between heterogeneous strata and various types of support elements (rock bolts, standing supports, yielding arches, and backfilling) in highly deformed systems. Compared with conventional calculation methods, numerical modeling can, for example, provide a detailed analysis of the fracture and deformation status of the rock strata surrounding a mining excavation. This produces a picture of the extent of the strata fracture zone and helps to determine the support measures required. To understand the complex mechanical phenomena within the rock mass with all its inhomogeneity and separation planes, it is necessary to develop and implement various constructive models in the form of mechanical equations, which can reproduce the stress-deformation behavior of the rock mass with sufficient accuracy.

DMT has used several numerical programs for solving different geomechanical questions for many years. For large-scale stress redistributions, the program GEDRU developed by DMT is used. For most planning cases, the program FLAC (two- and three-dimensional) is applied. In special cases, other programs like PFC, UDEC, or Ansys are used. The main focal point here is the application of the program FLAC to solve several geotechnical questions.

The German coal industry uses numerical models for a wide range of planning assignments, including:

- Warranty of long-term stability for underground openings like shafts, pit bottoms, or main roadways
- Investigation of the deformation behavior of temporary underground openings such as longwalls or gate roads
- Support dimensioning
- Investigation of mining impact (surface subsidence, damage)

The numerical modeling process essentially operates by reproducing all fundamental support elements (rock bolts, standing supports, roadside packs, injection material, etc.), along with a wide range of boundary conditions that apply to the working of coal seams in geological deposits (geology, depth, single- and multiple-seam mining, etc.). This paper uses different examples to present the range of application of numerical modeling techniques.

PREPARATION OF NUMERICAL MODELS

The preparation of geomechanical-numerical models for control of underground openings requires a multitude of information as initial parameters. This includes:

- Strata
    - Geology
    - Geotechnics
- Type and shape of the underground opening
    - Two-dimensional simulation (e.g., roadways)
    - Three-dimensional simulation (e.g., junctions)

[1]IMC-Montan Consulting GmbH, Essen, Germany.
[2]DMT GmbH & Co. KG, Mining Service, Essen, Germany.

- Rock stress
  - Primary rock stress
  - Secondary rock stress
- Rock support
  - Support elements
  - Support pattern
  - Support behavior

To create models that are as realistic as possible, initial information with a high accuracy and model calibration are required. To obtain this information, many investigations need to be conducted (Figure 1):

- Geologic and geotechnical core logs
- Rock mechanical lab tests
- Stress measurements and calculations
- Underground support tests
- Laboratory support tests for the support material as well as for the installed support elements (resined bolts including the rock mass)

The challenge is the modification of the lab parameters for use in the continuum-mechanical program FLAC. Due to the limitation of implementing every geotechnical element (e.g., joint) in a model, a reduction of these test results (e.g., rock strength) is required. Therefore, different empirical approaches and experiences are used. An additional point is a mostly limited amount or poor quality of initial parameters. For these reasons, calibration of the model is required. DMT uses two different methods for model calibration [Hucke et al. 2006]:

- Physical modeling of underground roadways (1:15 scale)
- Underground measurements and observation

The advantage of this extensive calibration process is that one gains a comprehensive picture of the complex deformation and fracture structure around the underground opening. The physical modeling technique especially permits a detailed view into the rock around an opening. Thus, it is possible to detect fractures in the rock as well as support failure (e.g., bolt failure).

Based on these considerable initial parameters and calibration capabilities, DMT has successfully developed numerical modeling as a reliable planning tool that can be used worldwide. The following examples describe the application of numerical models for different planning cases.

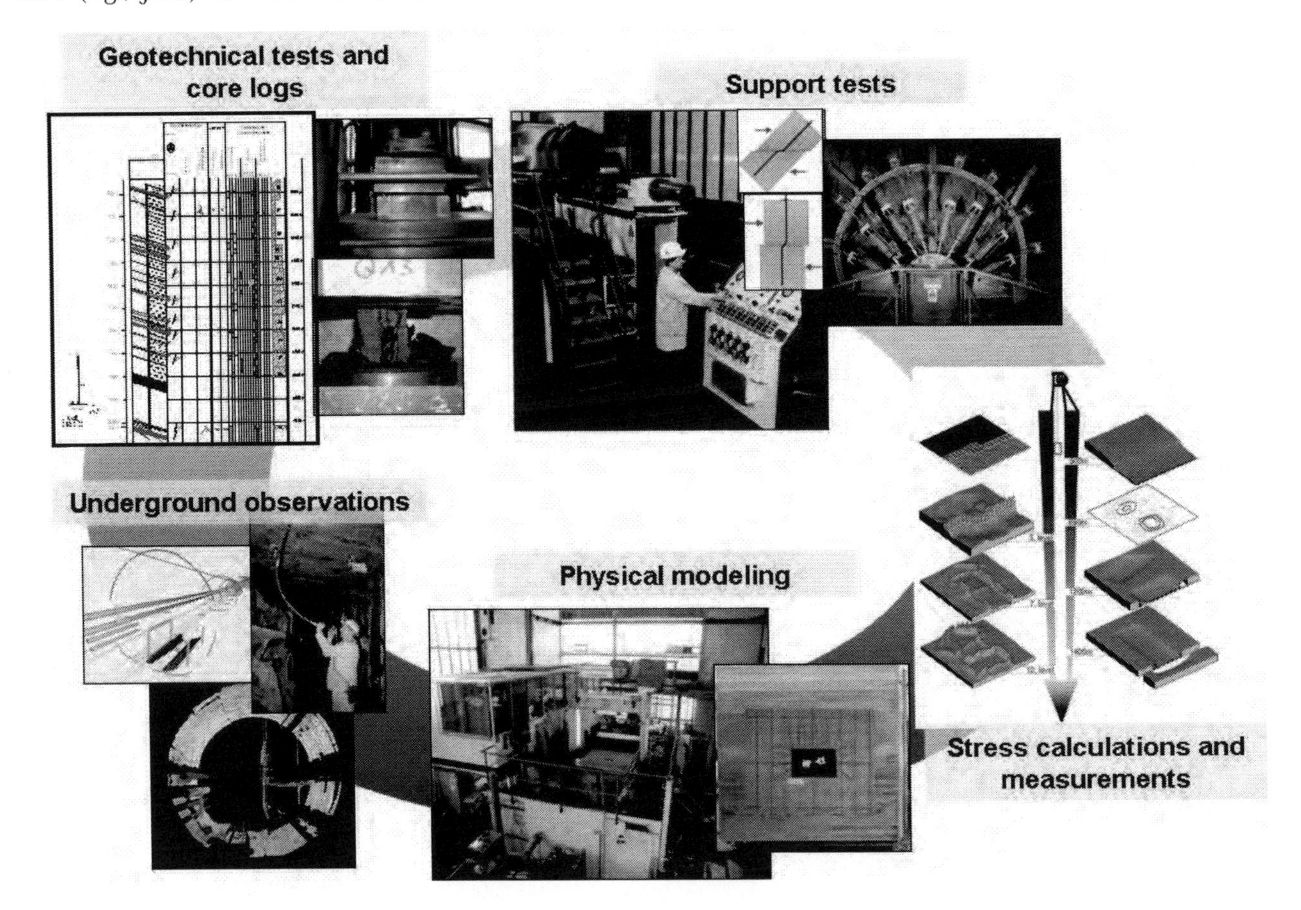

Figure 1.—Information required for calibrating a numerical model.

APPLICATIONS

This section will focus on the various ways in which underground openings are used and how this affects the numerical modeling process. This will be done using examples taken from different mining operations:

1. Gate road at average depth in brittle ground at a Ukrainian coal mine
2. Dimensioning of stable pillars to prevent spontaneous combustion in an Italian lignite mine
3. Investigation of rock burst risk for multiple-seam extraction in a Czech coal mine
4. Dimensioning of roof bolt support for roadways and junctions in a German gypsum mine
5. Dimensional design of shaft and pit bottom support systems at a depth of 1,200 m in a German coal mine
6. Investigation of surface subsidence as a result of extraction in German coal mines
7. Room-and-pillar design in very weak ground at a Russian iron ore mine
8. Planning of shape and driving sequence of a rise heading in a German coal mine

1. Ukrainian Coal Mine: Planning of Gate Roads

Investigation Focal Point

The investigation focused on the search for alternative roadway support designs combined with roadway shape variations with the key objective of substantially reducing the considerably high roadway deformations. A special feature of this case was the lower strength of the surrounding rock mass compared with the coal, which is the main reason for the high roadway deformation under these conditions.

Results

The numerical setup for the basic model depends on a visual evaluation of the in situ strata behavior. Figure 2 compares the numerical model with the actual deformation taking place in the roadway.

To improve the stability of the roadway and reduce its deformations, several variations were investigated based on this calibrated model. Two results are presented in Figure 3. The left image shows a roadway with backfill and yielding arch support, which could improve the roadway stability considerably. The right image shows a rock-bolted rectangular roadway, resulting in poor stability and high roadway deformation.

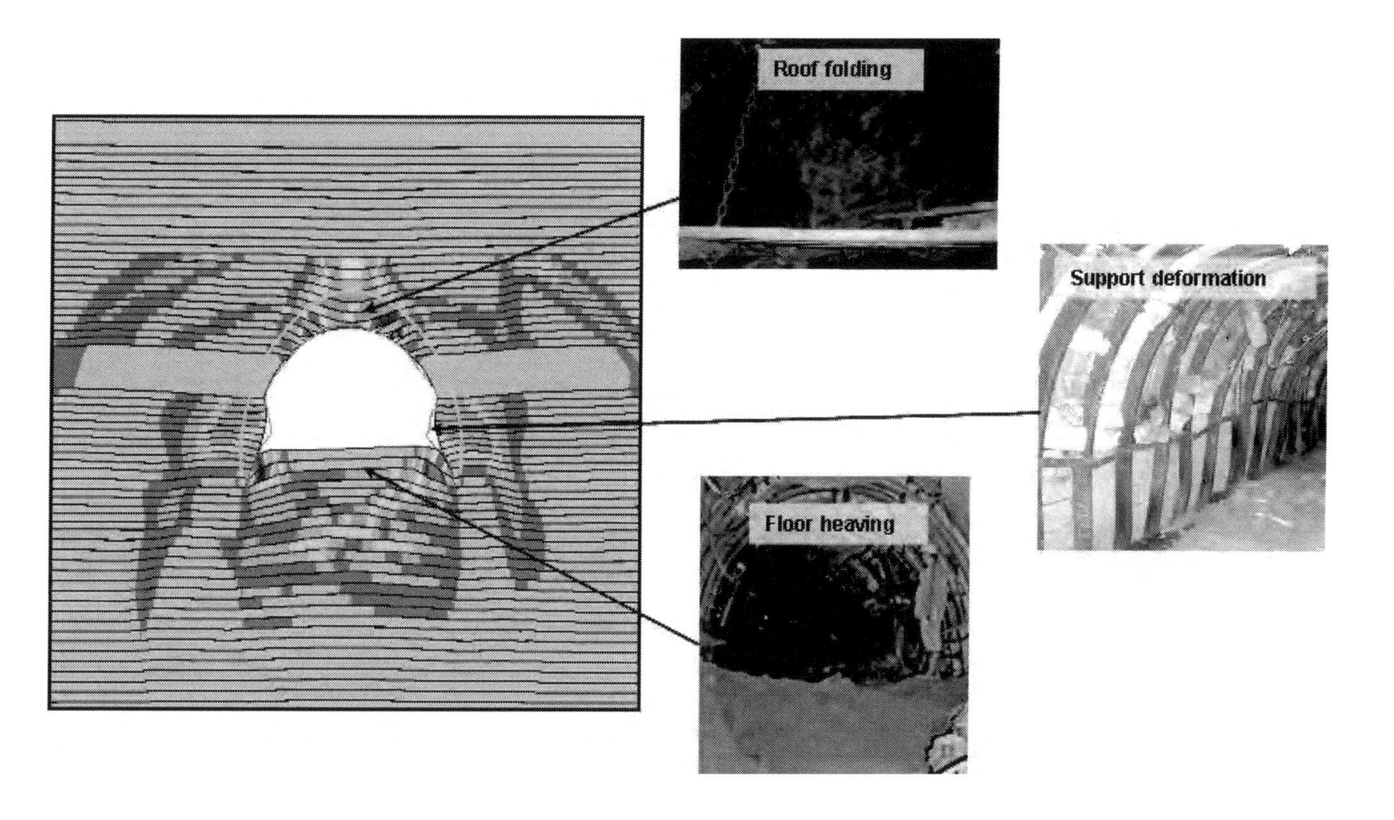

Figure 2.—Roadway deformation structures. The yellow line in the left image indicates the initial cross-section.

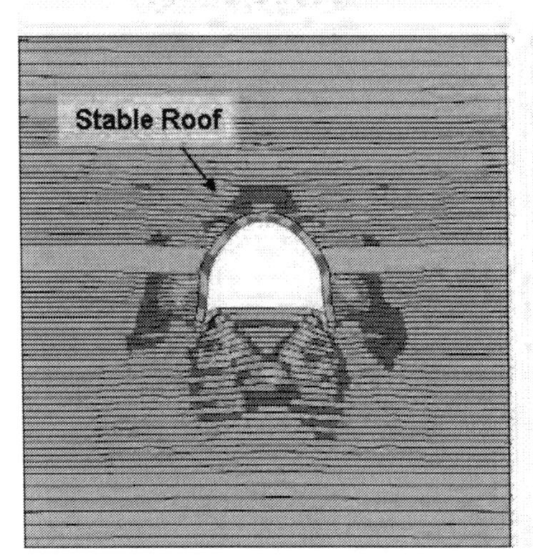

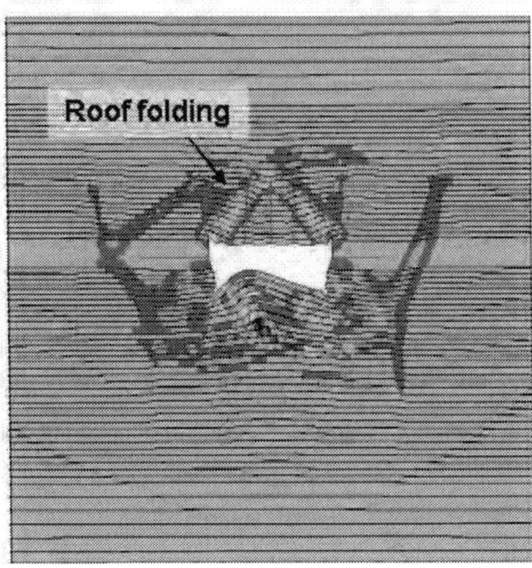

Figure 3.—Fracture zones for different investigation cases.

Outcome: A suggestion for improved roadway support was provided based on a "visual model calibration," taking into consideration the specific boundary conditions [Studeny and Wittenberg 2007].

2. Italian Lignite Mine: Pillar Design

Investigation Focal Point

The main objective was the dimensioning of stable pillars without fracturing to reduce the risk of spontaneous combustion of a high compact layer on the surface. A special concern was the improved calculation of pillar width with a stable, nonfractured core inside compared with simple empirical approaches (pillar width = 10% of depth) to optimize the utilization of the deposit.

Results

Based on underground observations, a calibrated model for one roadway was created first. In the second step, the model was extended to reproduce the planned situation including extraction and the second roadway. Therefore, several pillar widths were investigated. Figure 4 shows two cases: fractured pillar *(left)* and stable core in the pillar *(right)*.

The numerical results were compared with those of the Analysis of Longwall Pillar Stability (ALPS) program (Figure 5). The result shows very good agreement between these two calculation tools.

Outcome: A new approach for designing pillar width was developed with a reduction from 10% to 7.5% of the mining depth, resulting in higher utilization of the deposit.

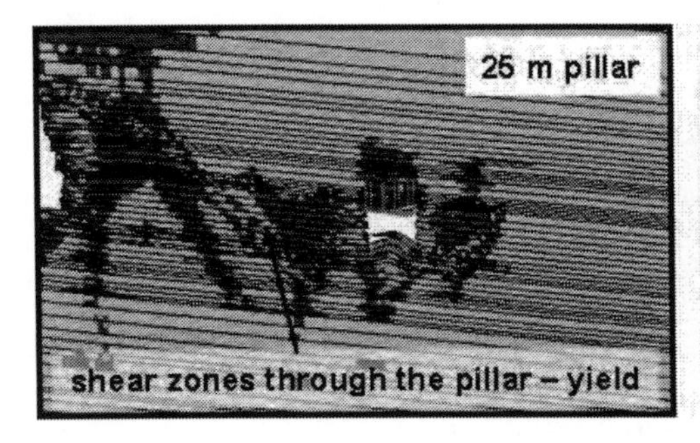

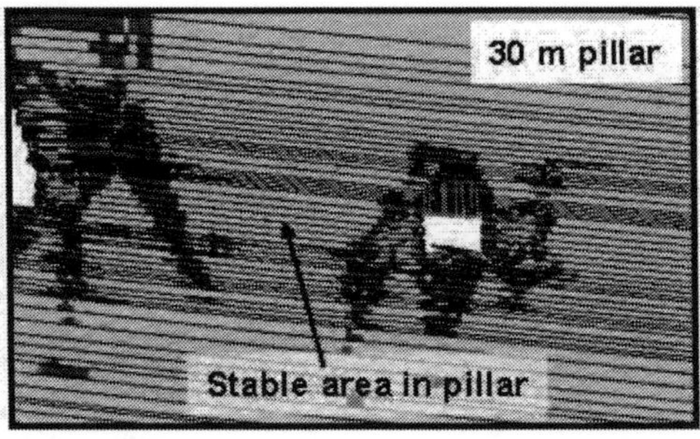

Figure 4.—Shear zones around two roadways for different pillar widths.

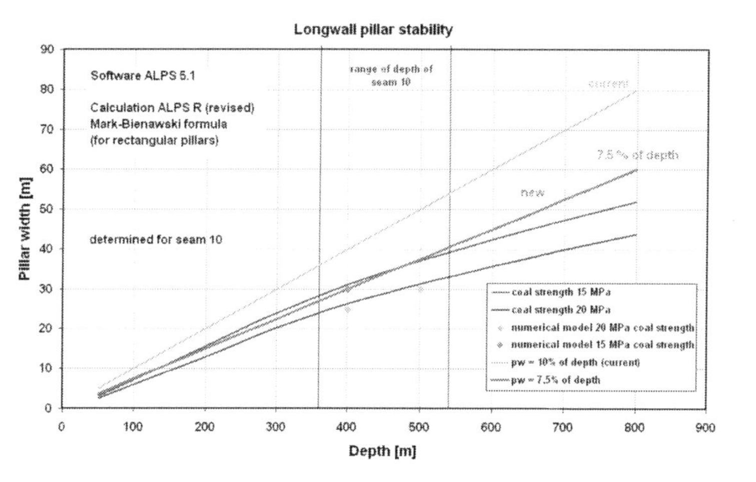

Figure 5.—Pillar width analysis: comparison between numerical model and ALPS calculation.

3. Czech Coal Mine: Rock Burst Prevention

Investigation Focal Point

The objective of the investigation was to analyze the stress situation in a mining seam in different areas to detect a potential rock burst risk in the seam, including the influence of multiple-seam extraction. Based on large-scale stress calculations with GEDRU for the complete mining area, small-scale FLAC models need to be determined. These models include mining in different seams as well as large-scale stress distribution as initial conditions for modeling.

Results

The current situation is shown in Figure 6 with a new panel next to the excavated area. During the mining process, rock bursts occur.

The red lines in Figure 6 depict a residual pillar overlying extracted areas. Based on this, large-scale stress analyses were conducted for the whole mining area to show potential rock burst risk areas. In Figure 7, two sections are presented: next to the extracted area near the tailgate *(left)* and near the main gate *(right)*. The high stresses around the extraction near the tailgate were an indicator of a rock burst risk. The stresses near the main gate are considerably lower.

Outcome: A map of potential rock burst risk areas was compiled, and measures for rock burst prevention were developed by rock burst experts based on these findings [Breitenstein et al. 2007; Baltz and Hucke 2008].

4. German Gypsum Mine: Support Design

Investigation Focal Point

The investigation focused on improving the roof stability of roadways and junctions extracted in layered strata to avoid cleavage of the roof as a result of thin clayey layers. The calculations were done with two-dimensional models for roadways and three-dimensional models for junctions.

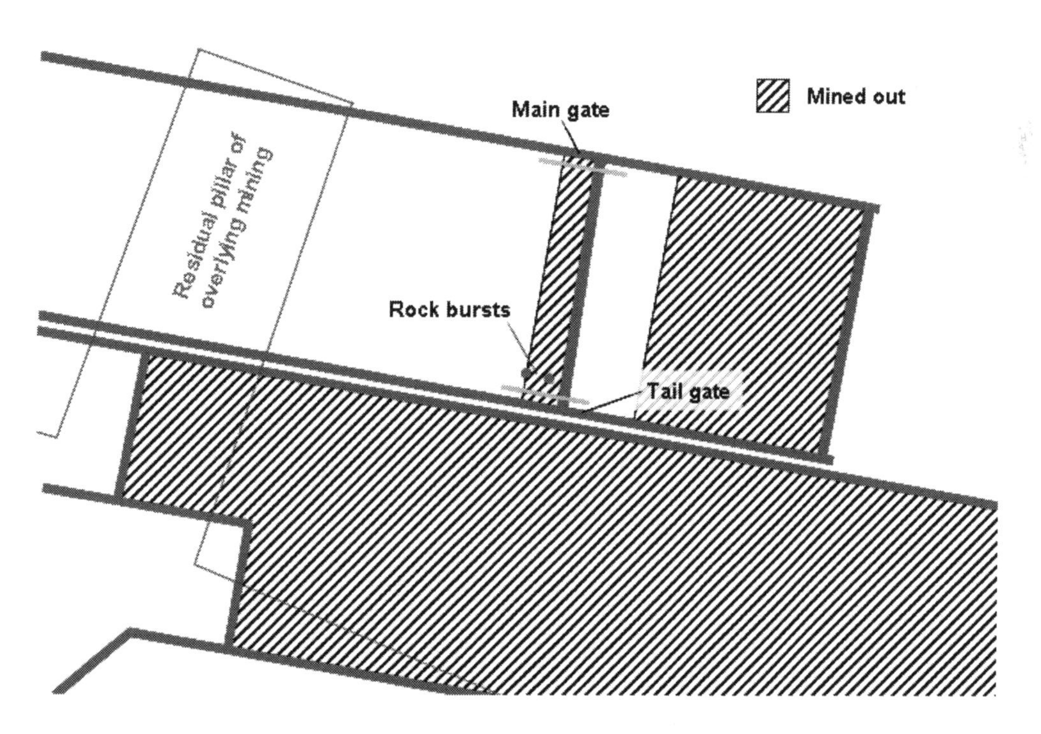

Figure 6.—Ground view of investigation area with marked model cross-sections (green lines).

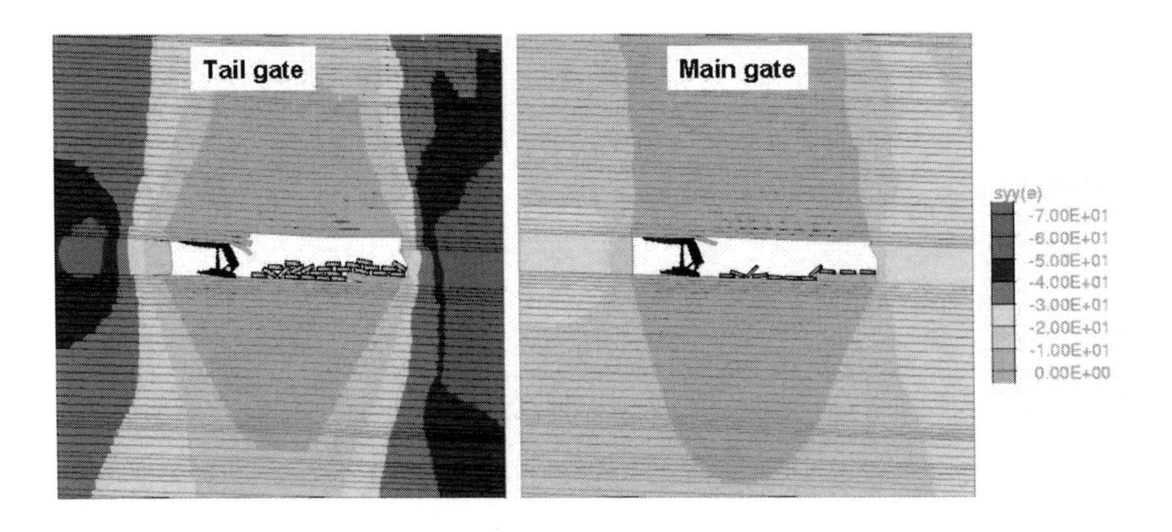

Figure 7.—Vertical stress during excavation for two investigation areas.

Results

The stability problems were the result of thin clayey layers in the roof that were very sensitive to moisture. Due to loosening at infiltration, the stability of the roof is dependent on the moisture of the clay. In particular, the junctions are endangered because of the large span. Therefore, several three-dimensional calculations with different layer strengths were conducted, which resulted in a support pattern with Swellex bolts to avoid cleavage of the roof (Figure 8).

Outcome: By combining the modeling results with empirical approaches, the support system could be optimized significantly for the roadways and junctions.

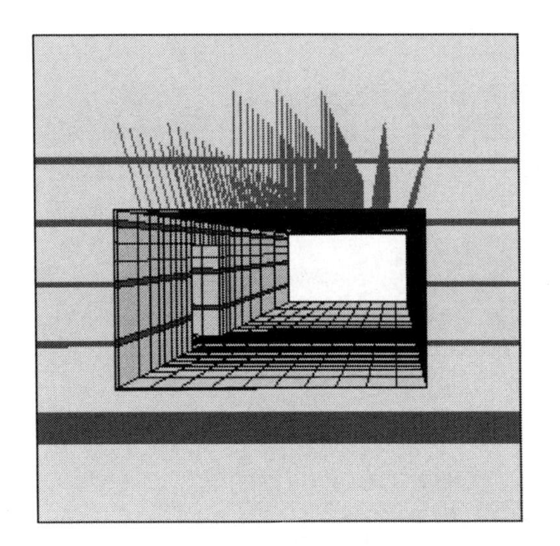

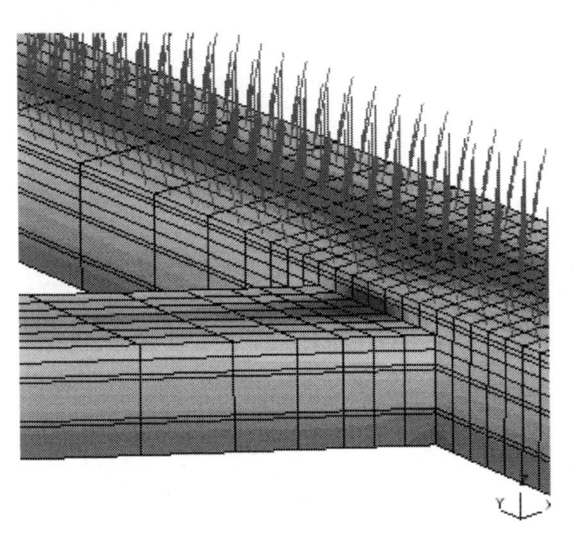

Figure 8.—Suggested support pattern for a junction.

5. German Coal Mine: Pit Bottom and Shaft Support Design

Investigation Focal Point

The main question was the stability of an installed rigid support of a pit bottom, with a cross-section of approximately 200 m^2 and a length of 15 m, due to changing stress conditions caused by drivage of a roadway toward the pit bottom. The challenge was to create a three-dimensional numerical model with a demanding geometry, layered strata, different stages of excavation, support installation, and fault implementation.

Results

To solve this task, a complex three-dimensional model was required. The geologic-geomechanical conditions and the design variations for the connection between the existing roadway system and the already built pit bottom placed a high demand on the model. Figure 9 shows a section of the model with the layered strata, the shaft, the built pit bottom, and first steps of roadway drivage. The roadway drivage was planned in two steps—the upper contour, then the floor.

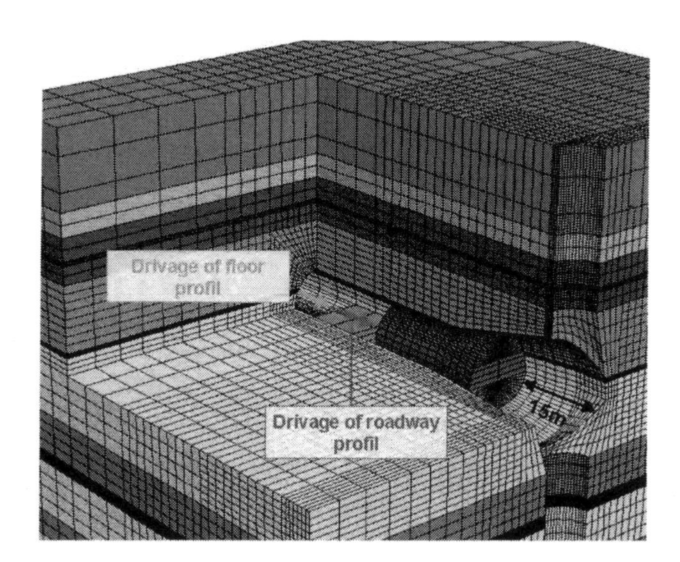

Figure 9.—Section of the numerical model with drivage sequence.

By means of modeling, a roadway support concept was developed for the convergence and breakthrough to the existing, rigidly supported pit bottom. Therefore, stress analyses were conducted to investigate the pit bottom support load for several variations of roadway support. Results showed that installing high-powered supports, 10 m in front of the breakthrough to the pit bottom, would protect the existing rigid pit bottom support considerably.

Outcome: Numerical modeling was used successfully to develop a construction concept for such a complex underground building.

6. German Coal Mines: Surface Subsidence

Investigation Focal Point

This project investigated the influence of different parameters such as extraction, advance rate, seam thickness, strata conditions, panel dimension, and multiple-seam mining on surface subsidence and subsidence rate. A major point, compared to generally used stochastic models for subsidence prediction, was the consideration of geologic-geomechanical parameters, as well as mining activities.

Results

To calibrate the numerical models, subsidence measurements at the surface of several German coal mines were used. Therefore, different situations such as single- and multiple-seam extraction, as well as single and multiple panels in one seam, were examined.

Figure 10 compares numerical calculations and measurements for an actual case. In a virgin mining area, one panel was extracted, then two more panels next to the first were mined simultaneously. Both the lower subsidence after mining the first panel (red line) and the higher subsidence after three extracted panels (blue line) could be reproduced with the same numerical model without changing any parameters, which is nearly impossible with common stochastic subsidence calculation.

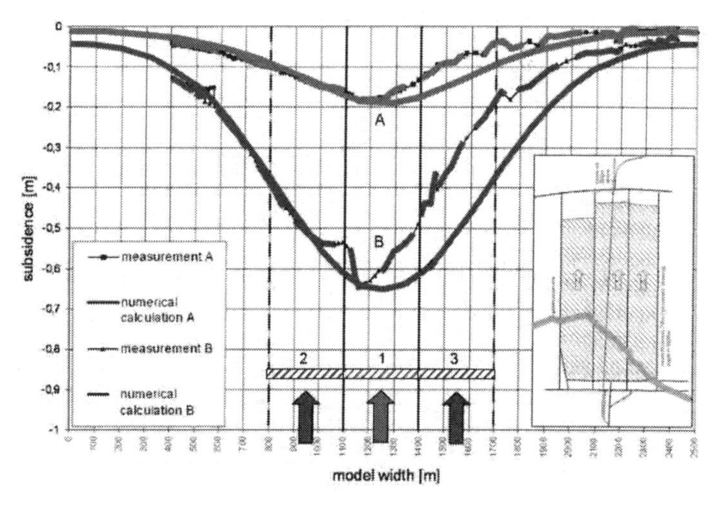

Figure 10.—Measured and calculated subsidence on the surface.

Outcome: A geomechanic-numerical model was created that can reproduce the surface subsidence for various mining situations, taking into account the geologic-geomechanical boundary conditions of the strata [Kook et al. 2008].

7. Russian Iron Ore Mine: Room-and-Pillar Design

Investigation Focal Point

The main objective was the planning of a room-and-pillar test mining layout in a weak rock mass, including room, pillar, and support design, to define the boundary conditions for complete extraction of an ore body. A special feature was the reproduction of the existing roadways with small deformations in a rock mass, which is characterized by behavior and strength similar to a soil, at a depth of approximately 600 m.

Results

The model calibration was based on underground observations in existing roadways in the mine. The iron ore rock mass was characterized by a very low strength and a high water sensitivity. To reproduce the observed roadway deformations, a strength of 1 MPa was used for the modeling. With this calibrated model (Figure 11, *left*), a multitude of variants were investigated to design the room (with or without support) and the pillar for a test mining area. A highly effective mining concept with large rooms was not possible because of stability problems with the room (Figure 11, *center*). Therefore, a concept with rock-bolted roadways and a pillar width twice the room width was recommended (Figure 11, *right*).

Outcome: A mining concept for a test mining layout in highly stressed, weak rock mass was provided to and implemented by the mining company.

8. German Coal Mine: Rise Heading Design

Investigation Focal Point

The focus of this investigation was the planning of the shape and driving sequence of a rise heading at great depth in a highly disturbed rock mass for a RAG mine in cooperation with RAG experts. A special feature was the implementation of different combined support systems (rock bolts, yielding arches, standing support, shield support) and partially combined roadway shapes (arch shape with rectangular extension roadway for shield support) in one model.

Results

Based on experience and empirical approaches, a model calibration for the basic roadways (purely arch-shaped and rectangular roadway) was conducted. These calibrated models were used in the investigation of different variants of rise heading drivage in RAG coal mines.

Results are shown in Figure 12. The left image shows an arch-shaped roadway with a rectangular extension in the mining direction. The middle and right images show a rectangular roadway also with an extension in the mining direction (to the left) with different support systems. The middle image has only standing support, while the right image has bolts and powered support. It was found that only a powerful support with shields and roof bolts could reduce the roof fractures and roadway deformation considerably. In the other cases, visible fractures in the roof could result in stability problems during mining operation.

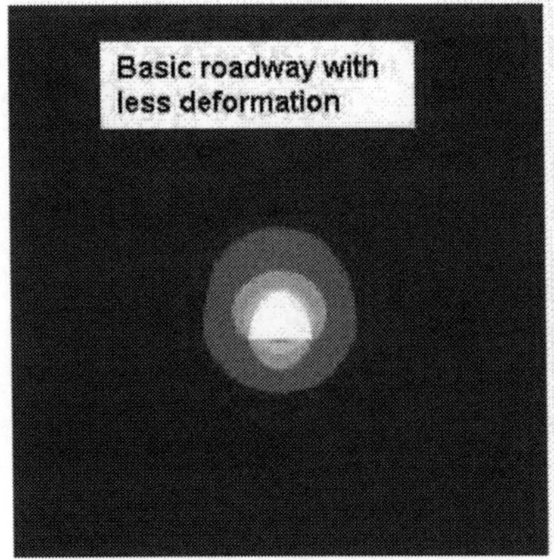

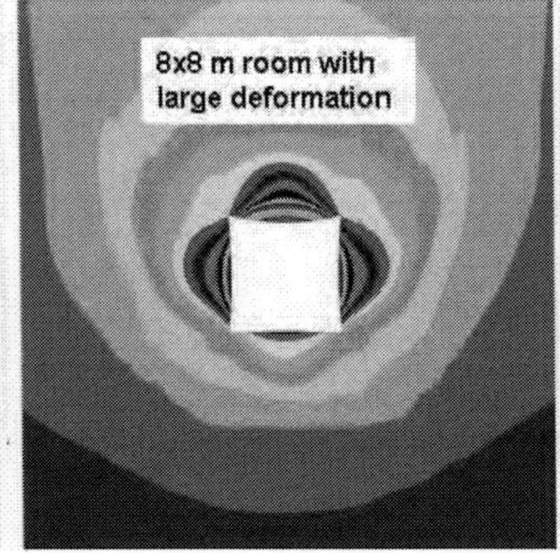

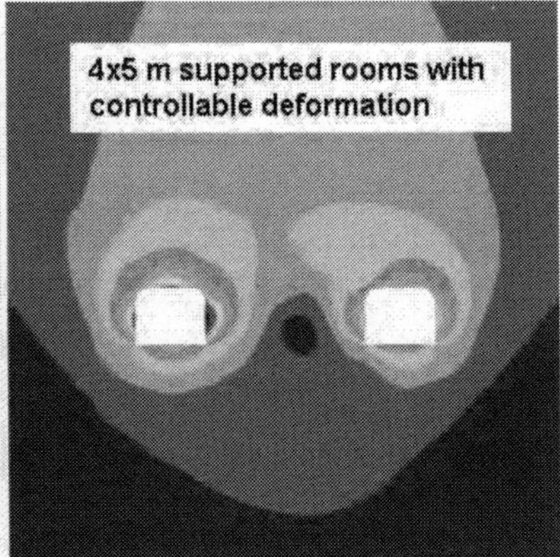

Figure 11.—Roadway and room deformation for different investigation cases.

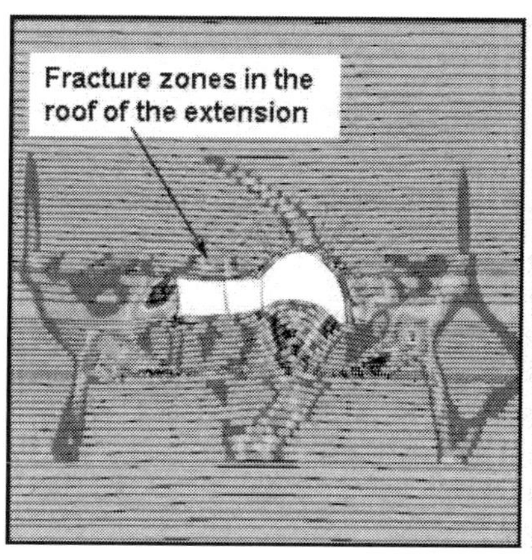

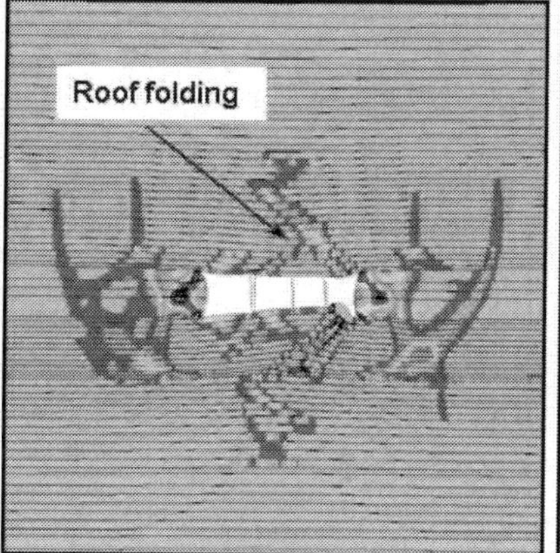

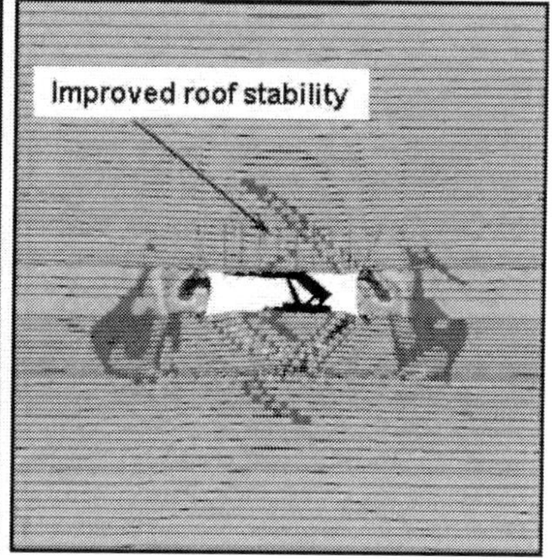

Figure 12.—Fracture zones around the rise heading for different investigation cases.

Outcome: Based on experience, plausible models were created to support the client (RAG) in choosing a rise heading design.

CONCLUSIONS

Based on extensive experience over many years in underground mining and physical modeling, DMT has accumulated a comprehensive understanding of geomechanical processes, especially with regard to the interaction of rock stress, rock mass structure, and support technology. This knowledge was implemented in geomechanical-numerical models and, due to continuous calibration and optimization, a high standard of modeling is assured.

As proven by various projects, this advanced modeling technique is suitable for solving a wide variety of geotechnical problems in underground openings in different kinds of deposits. This confirms that numerical modeling is a versatile tool in current mine planning and shows increasing importance for safe, economic, and trouble-free mining operation in the future.

REFERENCES

Baltz R, Hucke A [2008]. Rockburst prevention in the German coal industry. In: Peng SS, Tadolini SC, Mark C, Finfinger GL, Heasley KA, Khair AW, Luo Y, eds. Proceedings of the 27th International Conference on Ground Control in Mining. Morgantown, WV: West Virginia University, pp. 46–50.

Breitenstein K, Huwe HW, Baltz R [2007]. Protective seam extraction as a planning measure for the efficient mining of rockburst-prone coal seams. In: Proceedings of the Fourth International Symposium on High-Performance Mine Production (Aachen, Germany, May 30–31, 2007).

Hucke A, Studeny A, Ruppel U, Witthaus H [2006]. Advanced prediction methods for roadway behaviour by combining numerical simulation, physical modelling and in-situ monitoring. In: Peng SS, Mark C, Finfinger GL, Tadolini SC, Khair AW, Heasley KA, Luo Y, eds. Proceedings of the 25th International Conference on Ground Control in Mining. Morgantown, WV: West Virginia University, pp. 213–220.

Junker M [2006]. Gebirgsbeherrschung von Flözstricken (in German). Essen, Germany: Verlag Glückauf GmbH.

Kook J, Scior C, Fischer P, Hegemann M [2008]. Subsidence prediction for multiple seam extraction under consideration of time effects by the use of geomechanical numerical models. In: Peng SS, Tadolini SC, Mark C, Finfinger GL, Heasley KA, Khair AW, Luo Y, eds. Proceedings of the 27th International Conference on Ground Control in Mining. Morgantown, WV: West Virginia University.

Studeny A, Wittenberg D [2007]. Numerical modeling for roadway support systems: a comparison for single and multiple seam mining. In: Peng SS, Mark C, Finfinger GL, Tadolini SC, Khair AW, Heasley KA, Luo Y, eds. Proceedings of the 26th International Conference on Ground Control in Mining. Morgantown, WV: West Virginia University, pp. 42–48.

Printed in Great Britain
by Amazon